NATIONAL ECONOMIC DEVEI

GUIDELINES FOR THE MANAGEMENT OF MAJOR CONSTRUCTION PROJECTS

NEDC CONSTRUCTION INDUSTRY SECTOR GROUP

LONDON: HMSO

First published 1991

ISBN 0 11 701219 X

British Library Cataloguing in Publication Data
A CIP catalogue record for this book is available from the British Library.

HMSO publications are available from:

HMSO Publications Centre
(Mail and telephone orders only)
PO Box 276, London, SW8 5DT
Telephone orders 071-873 9090
General enquiries 071-873 0011
(queuing system in operation for both numbers)

HMSO Bookshops
49 High Holborn, London, WC1V 6HB 071-873 0011 (counter service only)
258 Broad Street, Birmingham, B1 2HE 021-643 3740
Southey House, 33 Wine Street, Bristol, BS1 2BQ (0272) 264306
9-21 Princess Street, Manchester, M60 8AS 061-834 7201
80 Chichester Street, Belfast, BT1 4JY (0232) 238451
71 Lothian Road, Edinburgh, EH3 9AZ 031-228 4181

HMSO's Accredited Agents
(see Yellow Pages)

and through good booksellers

CONTENTS

		Page
	Foreword *Sir Christopher Foster Chairman, NEDC CONSTRUCTION INDUSTRY SECTOR GROUP*	*5*
1	Introduction *Christopher Tayler SHELL UK*	*7*
2	Project Specification and Initiation *Peter Morris BOVIS*	*10*
3	Planning Consent *Steven Aubrey SHELL UK*	*18*
4	Project Finance *Malcolm Clark COUNTY NAT WEST*	*23*
5	Contract Forms and Strategy *Bob Moyses FRANKLIN AND ANDREWS*	*30*
6	Project Control and Organisation *Alan Wills FLUOR DANIEL*	*40*
7	Quality *Bryan Burden SHELL UK*	*52*
8	Health and Safety *Victor Coleman HEALTH & SAFETY EXECUTIVE*	*57*
9	Design and Engineering *John Cooke DAVY McKEE*	*68*
10	Procurement of Materials *Trevor Kent KING WILKINSON*	*73*
11	Management of Subcontractors *Richard Weekes M W KELLOGG*	*81*
12	Construction *John Ling LING MANAGEMENT CONSULTANTS*	*88*
13	Improvement and Measurement of Productivity and Progress *Ivor Williams NATIONAL ECONOMIC DEVELOPMENT OFFICE*	*102*
14	Industrial Relations *Sir Pat Lowry NATIONAL JOINT COUNCIL FOR THE ENGINEERING CONSTRUCTION INDUSTRY*	*111*
15	Commissioning *Trevor Lane FOSTER WHEELER ENERGY*	*115*

FOREWORD

Major construction projects are complex undertakings. They involve the rapid formation of a project organisation equivalent in size to a substantial business, in order to plan, design and construct the project and, equally, quickly dismantle the entire organisation. Good management is an essential requirement if such an undertaking is to be successfully accomplished. The National Economic Development Council's Construction Industry Sector Group is concerned with improving the performance and execution of such projects. The 'Guidelines for the Management of Major Construction Projects' is not a theoretical treatise; it is a solidly based book written by experts for practitioners. That is why as Chairman I commend these Guidelines to all concerned with their undertaking.

Sir Christopher Foster
CHAIRMAN, NEDC CONSTRUCTION INDUSTRY SECTOR GROUP

The NEDC Construction Industry Sector Group would like to thank the authors of these Guidelines and would like to state that the ideas contained in the book are those of the authors and not necessarily those of the Sector Group.

1 INTRODUCTION

CHRISTOPHER TAYLER
Shell UK Limited

These 'Guidelines for the Management of Major Construction Projects' have been produced by the National Economic Development Council's Construction Industry Sector Group. They are comprised of a series of contributions written mainly by practitioners for use by practitioners and are intended to be a set of principles and checks for the total execution of major construction projects. Although the Guidelines cater for and refer to industrial process type projects, the principles in the main are equally applicable to non-process projects including building and civil engineering. The main purpose of the Guidelines is to provide a basic reference for the strategic management of projects, covering subjects and issues which have the greatest significance and effect in determining performance and results. The Guidelines are presented in chronological sequence, typical of most projects, and within the 15 subjects included, the reader will find the more obvious sections, such as Design and Construction, as well as the less obvious — Organisation and Planning Consent. Besides providing a straightforward guide for the steps to be taken during project execution, the reader will also discover that the subject chapters contain those salient features which can significantly influence the outcome of projects. Together with appropriate views from the authors, the reader should be able to make sound judgements of the main interrelated issues.

Clearly each project will have its own constraints and peculiarities which may dictate different directions from those suggested. It is not possible to outline strategic principles for all situations, but the reader may often be in a position to influence or change some of the constraints, in which case the Guidelines should prove a useful document for attaining the optimum approach.

Project Management, a business in its own right, is an old profession going back to Noah and beyond. Today it is of ever increasing importance as projects and their requirements become more complex. It is a vital ingredient of our society and economy. A well-managed project will not only be completed quickly, but will be cheaper, safer, of higher quality and therefore generally a better return on investment. That is the main aim of this publication. The salient aspects contained within should be clear enough to enable the framework of Project Management to be correctly set up and thereby avoid the pitfalls so often encountered.

However, this book is clearly not an encyclopedia and working details will be required from the specialists involved within each area to translate the concepts into realisable success with the hallmarks of quality. In managing a major project there are many aspects, other than those included in the subject headings of these Guidelines, which themselves also play significant roles. Nevertheless, one thing is certain — each and every project is different and will have its own problems, some unforseen and in some cases extensive. If the project is generally well set up and managed, these problems will be much easier to handle. However, to deal effectively with them, there are some essential ingredients such as: the strength, calibre and leadership of the project manager; his ability to be effective, determined and enthusiastic, energetic and decisive, without resorting to bureaucracy; the co-operation of all working on the project, whether client, managing contractor, erection contractor or workforce; and the proper sequencing of events, such as completing and freezing designs before proceeding with construction. Responsibility, too, plays a large part and there can be no satisfactory substitute for an organisation which focuses on single point responsibility. This not only concentrates the mind but also produces positive results, promotes job satisfaction, eliminates confusion, and forms a sound platform for good communications and understanding.

In setting up the basic elements, conflicts of interest can and do become a major stumbling block. No matter how good all the parameters, if there are major opposing forces at work, no project can be a success. These conflicts may take various forms, but undoubtedly financial aspects dominate forcibly throughout the whole project duration and have caused many significant and bitter disappointments. Ethics, too, form an important but not unrelated element. Proper codes of conduct and procedures are essential to ensure ethical behaviour. With these, the project will command respect and confidence from all parties. Obtaining the auditors' support for procedures before crossing the starting line will save many wasted hours of agonising post mortems.

Another area which frequently gives rise to disagreement and conflict is that of project progress, notably when related to construction. There is no substitute for separately measuring physical and financial progress. Although these elements are related, they rely on judgements and the predictions of each may vary significantly. Moreover, this whole aspect of project management and forecasting is based on judgement helped by systems and measurement. No matter how good the systems and procedures, Project Management is not an exact science and it is the element of judgement, together with realistic and frequent measurement, which will provide balanced and accurate trending and predictions and avoid unpleasant surprises.

All that has gone before is of little use unless the project has the support of the community. Public relations and community relations are vital ingredients. Following hard on the heels of these is resourcing. Obtaining the appropriate skills and tradesmen is clearly fundamental, but indications today are that this area will become a more significant problem in the years ahead. Willingness to change and adapt together with an investment in both training and innovation will be requirements for the success of the major projects of the future.

2 PROJECT SPECIFICATION AND INITIATION

PETER MORRIS
Director, Special Projects, Bovis Limited,

The way one sets up a project largely determines how successful it will be. We have known for generations that time spent at an early stage thinking through a problem — planning a project — is time well spent. Yet we consistently spend too little time in the project's early stages and we generally fail to consider adequately the range of issues which can subsequently cause problems.

From the lessons of success and failure of several dozen projects, both big and small, a model has been developed of those factors which must be achieved if a project is to be successful [Ref 1]. The crucial point is that all these items must be considered from the outset if the project's chances of success are to be optimised.

The model is shown in Figure 2.1. Briefly, its logic is as follows:

First, the project's objectives, technical base and general strategic planning need to be adequately considered and developed, and the design firmly managed in line with its strategic plans.

Second, the project's definition both affects and is affected by changes in external factors (such as politics, community views, and economic and geophysical conditions, availability of financing, and project duration). Therefore this interaction must be managed actively and well. (Many of these interactions operate, of course, through the forecast performance of the products that the project will deliver once completed).

Third, both the project's specification, its interaction with these external, financial and other matters, and its implementation will be much harder to manage and quite possibly damagingly prejudiced if the attitudes of the parties essential to its success are not positive and supportive.

Fourth, realisation of the project as it is defined, developed, built and tested involves:

- ▲ the crucial skill of moving the project at the appropriate pace through its development cycle;
- ▲ finding the right balance between the owner, as operator, and the project implementation team;
- ▲ having contracts which reflect these aims, which are motivational,

and which appropriately reflect the risks posed by the projects and the ability of the parties to bear these risks;

- ▲ establishing checks and balances between the enthusiasm and drive of the project implementation team and the proper conservatism of its owners/sponsors;
- ▲ treating project personnel as team members, with great emphasis being put on active communication and the handling of conflict in a productive manner;
- ▲ Having the right tools for project planning, control, reporting and quality assurance.

Figure 2.1 *The Management of Projects*

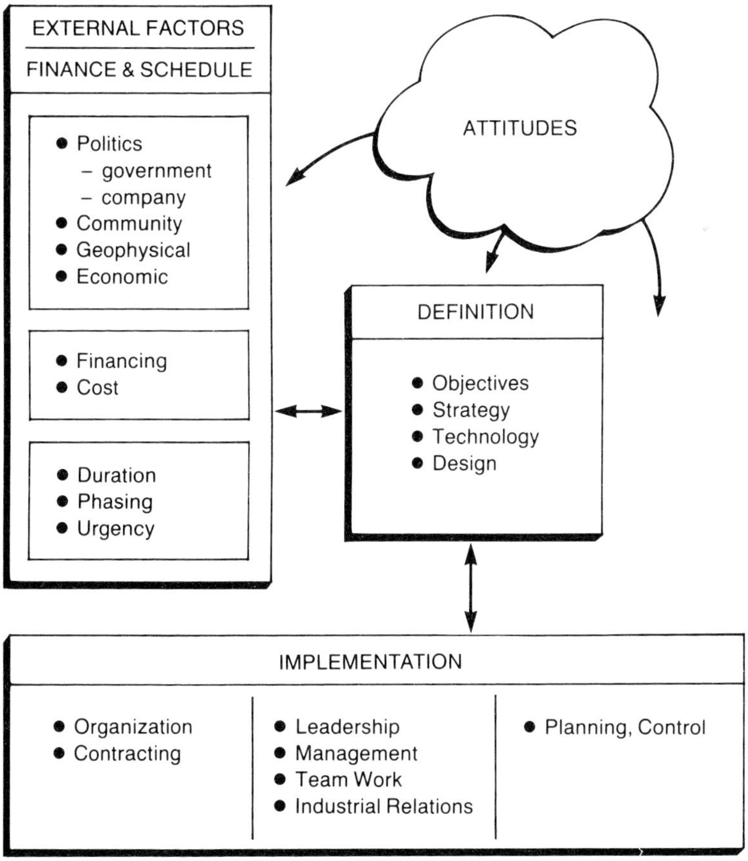

EXTERNAL FACTORS

FINANCE & SCHEDULE

- Politics
  - government
  - company
- Community
- Geophysical
- Economic

- Financing
- Cost

- Duration
- Phasing
- Urgency

ATTITUDES

DEFINITION

- Objectives
- Strategy
- Technology
- Design

IMPLEMENTATION

- Organization
- Contracting

- Leadership
- Management
- Team Work
- Industrial Relations

- Planning, Control

PROJECT SPECIFICATION

The extent to which the project's objectives are unclear, complex, do not mesh with longer-term strategies, are not communicated clearly, or are not agreed, will compromise the chances of project success. (Barry Turner examined various recent catastrophic accidents and found this factor to be of overriding importance [Ref 2]). The project should be defined comprehensively in terms of its type, location, energy and raw materials supply, transportation, product markets etc.

Strategies for the attainment of the project objectives should be developed in as comprehensive a manner as possible right from the outset. This means that at the pre-feasibility concept and feasibility stage industrial relations, contracting, communications, organisation, and systems aspects, for example, should all be considered, even if not elaborated upon, as well as the technical, financial, schedule and planning issues. Community support should be garnered.

There is abundant evidence that technical problems are a frequent source of problems in projects [Ref 3]. The development of the design criteria and the technical element of the project should be carefully defined. The design standards will affect both the difficulty of construction and the operating characteristics of the project. Maintainability and reliability should be determined. The rate of technology change in all relevant systems and sub-systems should be examined. The technical risk in particular should be assessed. Since theoretically no design is ever entirely complete — technology is always progressing — a central challenge typically is the resolution of the conflicting needs for meeting the schedule against the desire for continual improvement of the technical base. The orderly progressing of the project's design and its technical basis, with strict control of any proposed technical changes, must be a core element of modern project management. Thus, take care to assess the risks of prototypes or of new technologies, and to freeze the design before moving into implementation.

EXTERNAL FACTORS, FINANCE AND SCHEDULE

Several external factors may be identified but the project's political context, its relationship with the local community, the general economic environment, and its location and the geophysical conditions in which it is set are the most important.

External Factors

Most projects raise political issues of some sort and hence require political support: moral, regulatory, and sometimes even financial. National transportation projects, R&D programmes and many energy projects, for example, are subject to political decision and constraint.

Until the late '80s, for example, the civil nuclear power business had been heavily pushed politically. Third World development projects are especially prone to political influence. Even where the public sector is supposedly liberated to the private, as in Build-Own-Operate projects, political guidance and guarantees, as well as encouragement, are generally needed.

Do non-major projects also need to be conscious of the political dimension? Absolutely! Even small projects live under regulatory and economic conditions directly influenced by politicians; intra-organisationally too the project manager must secure 'political' support for his project.

The important lesson therefore is that these political issues must be considered at the outset of the project. The people and procedures that are to work on the project must be attuned to the political issues and be ready to address them. To be successful, project managers must manage upwards and outwards, as well as downwards and sideways. The project manager should court the politicians, helping allies by providing them with the information they need to champion the programme.

Though environmental matters have been affecting project implementation since the 1960s at least (the third London airport, Concorde, motorways etc), it is only during the 1980s that it has become more formally recognised as a project management issue. These days, most project staff realise that they must find a way of involving the community positively in the development of their project [Ref 4].

Establishing the support of the local community is particularly important. This will of course be difficult in situations where the project will have a direct environmental impact upon them. The balance may possibly be redressed by emphasising the benefits that the project will bring to the community at large.

Changes in economic circumstances affect both the cost of the project's inputs and the economic viability of its outputs. The big difference today compared with 20 years ago is that then we assumed conditions would not vary too much in the future; now, after the economic dislocation of the 1970s and '80s, the need for a cautious approach is required and more acceptable. Market surveillance is an essential part of the project's management in order to determine the viability of the project economics. In the area of cost-benefit analysis, NPV, discounting and other appraisal techniques, practice has moved forward considerably over the last few years. Externalities and longer term social factors are now recognised as important variables which can dramatically affect the attractiveness of a project. The basic project appraisal techniques of the 1960s have now been largely superseded by a broader set of economic and financial tools arrayed, in the community context, under the Environmental Impact Analysis (EIA) procedure.

Initially resisted by many in the project community, the great value of the EIA process is that it allows both consultation and dialogue between developers, the community, regulators and others, and forces time to be spent at the 'front end' in examining options and ensuring the project appears viable. Through these twin benefits, the likelihood of community opposition and unforeseen external shocks arising is diminished, and project developers are forced to spend time planning at the early stages of the project — the area which they have traditionally rushed, despite the obvious dangers.

Finance

During the 1980s there was a decisive shift away from public sector funding to the private sector. There is a belief that projects built under private sector funding inevitably demonstrate better financial disciplines. Where projects are built and financed by a well-managed private sector company this may be so, but good performance is usually a function of the way that the project is managed rather than how it is financed. What is required is funding realism. The best way to get this is by getting all parties to accept some risk and to make them face up to this squarely by undertaking a thorough risk assessment. Full risk analysis of the type done for limited recourse project financing for example invariably leads to better set-up projects and should therefore be built in to the project specification process. [Ref 5].

The raising of the finance required for the Channel Tunnel from the capital markets is a classic illustration of how all the elements shown in Figure 2.1 interact, in this case around the question of finance. To raise the finance required, certain technical work had to be done, planning approvals obtained, contracts signed, political uncertainties removed, and so on. Since the project was raising most of its funding externally, there was a significant amount of 'bootstrapping' required: the tasks could only be accomplished if some money was already raised. Actions had to be taken by a certain time or the money would run out. (Since there were only limited funds available only a limited amount of engineering could be performed — a constraint which was to cause considerable difficulties later when further engineering works, amongst other things, led to the project's cost estimates having to be raised 40 per cent to £7.5bn). Further, a key parameter of the project's viability was the likelihood of its slippage during construction. A slippage of three to six months meant not just increased financing charges but the lost revenue of a summer season of tourist traffic. The Channel Tunnel thus demonstrates also the significance of managing a project's schedule and of how its timing interrelates with its other dimensions.

Schedule

Determining the overall timing of the enterprise is crucial to calculating its risks and the dynamics of its implementation and management. How much time one has available for each of the basic stages of the project, together with the amount and difficulty of the work to be accomplished in those phases, heavily influences the nature of the task to be managed.

In setting up the project, therefore, the project manager will spend considerable effort ensuring that the right proportions of time are spent within the overall duration. Milestone scheduling of the project at the earliest stage is crucial. It is particularly important that none of the development stages for the project be rushed or glossed over — a fault which has caused many project catastrophes in the past. A degree of urgency should be built into the project, but too much may create instability.

Avoid implementation commencement before technology development and testing are complete. This situation is known as a 'concurrency' and though sometimes done quite deliberately (to get a project completed under exceptionally urgent conditions) it often brings major problems in redesign and reworking.

Concurrency is now increasingly synonymous with 'Fast Track', that is, building before basic design is complete. If faced with this, be under no illusion as to the risk. Analyse the risk rigorously, element by element, milestone phase by milestone phase. 'Fast Build' is now being used to distinguish a different form of design and construction overlap — where the basic design is completed but the work packages are priced, programmed, designed and built sequentially, within overall design parameters, with strict change (configuration) control being exercised throughout. With this 'Fast Build' situation, the design is secure and the risks are much less.

Each of the three areas — external factors, finance, and duration — is affected by and affects the viability of the Project Definition. They must all be managed by the project executive, and they must then be implemented through the project's life cycle.

ATTITUDES

Implementation can only be achieved effectively if the proper attitudes exist on the project. Unless there is a major commitment towards making the project a success, unless the motivation of everyone working on the project is high, and unless attitudes are supportive and positive, the chances of success are substantially diminished.

It is particularly important that there be commitment and support at the top; without it the project is probably doomed. (The study of the

Advanced Passenger Train in 'The Anatomy of Major Projects' illustrates this clearly [Ref 6].) But while commitment is important, it must be commitment to viable ends. The project should receive thorough and critical examination at the specification stage. Once authorised, it should receive full support subject to frank and objective reviews as it develops.

IMPLEMENTATION

Project management has in the past been primarily concerned with the process of implementation. This implies that developing the definition of the project is somehow not something which is the concern of the project's management.

The key point of this chapter therefore is not only that the setting up of the project must be actively managed, but that the initiation process must consider all those factors which might prejudice its success. These are not just technical matters, but economic and finance, ecological, political and community factors, as well as implementation issues. Subsequent chapters deal with more down-stream implementation matters in detail. This chapter has simply made the point that implementation begins from the very earliest stages of project specification and initiation.

REFERENCES

1. Morris, P. W. G., *The Management of Projects — Lessons From The First 50 Years of Modern Project Management* forthcoming.
2. Turner B., *Man-made Disasters, Taylor & Francis, London 1979.*
3. Many studies have attested to the importance of technical uncertainty. To the likelihood of overruns for example: Harman, A. J. assisted by Henrichsen, A., *A methodology for cost factor comparison and prediction*, Rand Corporation, R-6269-ARPA, Santa Monica, California, August 1970; Large, J. P., *Bias in initial cost estimates : how low estimates can increase the cost of acquiring weapon systems*, Rand Corporation, R-1467-PA & E, Santa Monica, California, July 1974; Marschak, T., Glennan, T. K. and Summers, R., *Strategy for R & D: Studies in the Micro-economics of Development*, Springer-Verlag, New York, 1967; Marshall, A. W. and Meckling, W. H. *Predictability of the costs, time and success of development*, Rand Corporation, P-1821, Santa Monica, California, December 1959; Peck, M. J. and Scherer, F. M., *The Weapons Acquisition Process : and Economic Analysis*, Harvard University Press, Cambridge Massachusetts, 1962; Perry, R. L., Smith, G. K., Harman, A. J. and Henrichsen, S., *System acquisition strategies*, Rand Corporation, R-733-PR/ARPA, Santa Monica, California, June 1971; Perry, R. L. DiSalvo, D.,

Hall, G. R., Harman, A. L., Levenson, G. S., Smith, G. K. and Stucker, J. P., *System acquisition experience*, Rand Corporation RM-6072-PR, Santa Monica, California, November 1969.
4. Stringer, J., *Planning and Inquiry Process*, MPA Technical Paper No. 6, Templeton College, September 1988.
5. Nevitt, P., *Project Finance*, Euromoney, London 1987.
6. Morris, P. W. G., Hough G. H., *The Anatomy of Major Projects*, John Wiley and Sons, Chichester 1987.

3 PLANNING CONSENT

STEPHEN J AUBREY
Shell UK Limited

'When will we be able to start work?' This is a question crucial to project planning and the critical path, but before construction can start, the necessary consents and authorisations must be in place. However, the time taken to get a planning application through the system is probably more unpredictable than any other element of the project. How many construction starts have been postponed because approvals have not yet been received?

> 'By simplifying the (planning) system and improving its efficiency, whilst ensuring that effective control can be maintained where it is warranted, the Government look to strike a balance between the needs of development and the interests of conservation'. (DTI White Paper 11/88 — Releasing Enterprise).

> 'Development should be prevented or restricted only where this serves a clear planning purpose and the economic effects have been taken into account. Development control must avoid placing unjustifiable obstacles in the way of any development especially if it is for industry, commerce, housing or any other purpose relevant to economic prosperity'. (Planning Policy Guidance Note No: 4 01/88 — Industrial and Commercial Development and Small Firms).

Yet the planning system is a complex bureaucracy and there are two issues, which more than any others, cause anguish to the project manager/construction engineer: timing and control, ie the uncertainty of when decisions may be reached and approvals given; and the lack of control over the numerous interested parties involved in those processes.

Although the average time for deciding planning applications is less than eight weeks, and Appeals on average are dealt with in three to four months, major construction projects will normally take far longer than these periods. Moreover there can be striking variations in the times which individual local planning authorities take to make decisions, and some will be more disposed to refuse permission than others. Broadly speaking, one will more readily get consent for major projects in areas of long standing economic or employment problems, whereas in many more prosperous parts of the country, particularly in the South East, Green Belt and other similar areas, limitations may impose severe constraints.

Control is fundamental to project management, of costs, design, specification, materials, timing etc. However during the Planning processes, one is almost entirely in the hands of local authority officials, politicians and statutory procedures whose inertia causes, if nothing else, frustration to the project manager.

So where are the pitfalls in the Planning process and what steps can be taken to avoid or minimise them? This chapter looks at the various areas that need to be considered but does not address those other external controls and restraints that run alongside Planning, such as building regulations, Health and Safety Executive, Factory Inspectorate, hazardous substances and other works requiring Registrations, pollution, water, sewerage, waste disposal, Fire Certification, gas, electricity, highways and many more. All need to be addressed, and all can cause delay where local authority consultation is involved.

There is much useful work that can be done in advance of making a planning application and the 'planning' of these processes should be a fundamental part of the construction project.

Apart from the physical aspects of the land on which one is proposing to build — soil and site surveys, levels, contamination, etc — to whom does the land belong? Are there constraints in the legal title of the land? Do previous owners or other parties have a say in the development proposals? Are there landlords or tenants to consider? What other individuals or bodies are going to be involved in the Planning process? Setting out the framework, and perhaps early consultation, will minimise uncertainty and help to define a much more realistic schedule. If the application ends up at Appeal, this early planning should already have identified the problem areas. Consider the introduction of planning consultants. They may well be required ultimately and, if this seems likely, bring them in now.

Does the project require planning consent? It is not only building works that need permission. 'Engineering' and 'mining' operations are also included in the definition of development, as is 'change of use'. It is best to assume that planning permission is required, since Permitted Development Rights in the General Development and Use Classes Order will probably not be of significant help. Within the definitions of Permitted Development, there are of course many instances of deemed planning consent. For instance, Local or Private Acts of Parliament or Orders approved by both Houses of Parliament, though rare, may deem a planning permission. Major construction projects, usually of more than local interest such as new railway links, river crossings, etc often follow this route. A cross-country pipeline authorised by the Secretary of State is a similar development outside the usual local planning system. Such procedures are no less time consuming than the usual Planning processes, and often more so. Development by the Crown on Crown land does not require planning permission, but it is a long standing convention that the Crown should seek an informal clearance from the local planning authority on similar lines to the submission of a planning application.

Does your proposal involve contentious issues? Have previous projects gone ahead without adherence to the necessary consents or conditions? Are you supported by Structure/Development Plans and/or Local Plans — the statutory documents that set out planning policy for an area. DoE Planning Policy Guidance Notes may help to explain a local authority's expected response to your application.

Will you require an Environmental Impact Assessment? These are now mandatory for certain types of development, to accompany an application. For relevant projects, this will describe the significant effects that the proposed development would have on people, flora and fauna, soil, water, air, climate, landscape, 'material assets' and the 'cultural heritage'. A 'relevant project' will be one which will have a significant impact on the environment, by virtue of its nature, size or location, usually:

(a) where a major project has more than local importance,

(b) where a project is proposed for a particularly sensitive or vulnerable location, or

(c) where a project has unusually complex and potentially adverse environmental effects.

The preparation of Environmental Statements can be very time consuming, as well as costly. Consultation with Environmental Health Officers may provide useful guidance on the content and presentation.

There are still many other issues to consider, including: structure and local plans, enterprise zones, simplified planning zones, urban development areas/corporation, development/assisted areas, listed buildings and historic monuments, archaeology, green belts, conservation areas, sites of special scientific interest, national parks, nature reserves, forestry, areas of outstanding natural beauty, tree preservation orders, demolitions, building lines, heights of buildings, use of materials, employment generation, public footpaths and rights of way, car-parking requirements, construction traffic, delivery traffic movement and congestion, parking, unloading and turning of vehicles, outstanding enforcement proceedings, etc.

Remember the local authority has to consult numerous bodies about your application — county/parish councils, statutory undertakers, etc. This takes time and may raise questions or objections.

Prepare the ground with your local planning officer. It is rarely advisable to submit a 'blind' application. The scale or nature of the project may determine the level of consultation. Pre-application discussions with the appropriate officer can avoid many problems later on. He may not commit himself, or his authority, but he will offer valuable guidance. He can charge you however for a consultation as indeed he will charge fees for the application. It may be possible to discuss, and perhaps agree, the conditions that are likely to be attached to a consent if given. Similarly the scale or nature of the project may promote the subject of planning gain or Section 106 (exs.52) Agreements. These are in effect the local authority's requirements of a developer, in exchange for their consent, that cannot adequately be dealt with in the conditions of the permission. They have to be reasonable and relate to the development but may include highway modifications, provision of open space, etc.

How is the application going to be dealt with? When are the committee dates? What are the lead times? Who are the consulted parties? Will the planning officer have delegated powers? Beware Summer holidays. There are often long gaps between committees during July, August and September.

What about notices concerning the application — notices to owners, tenants; newspaper advertisements; advertisements on land, etc. Do make sure that your application is not rendered invalid because you have failed to observe certain procedures — the result will be delay.

Lastly, make sure the application is correctly put together with the right number of copies, plans and detail, leaving no ambiguities or detail missing that will force you to start the process over again. It may be advisable to accompany the application with a covering letter describing the proposed development, the reasons for it and the approach taken. Glossy presentations and handouts to officials and politicians do not always ensure success, but the provision of supporting information may be of great help. A minefield? It seems that way. Do you have the expertise to steer a clear path through it? Consider again the inclusion of a planning consultant in the project team. When submitting the planning application, should one request outline consent or full-detailed permission? An outline consent to establish, in principle, whether the proposed development is acceptable may be better strategy in the long run.

If you have followed the above, you should have a good idea as to whether you are likely to get a consent or a refusal, what the conditions might be, whether you are going to appeal and when you might expect all of this to happen. If you expect a consent on a certain committee date, add on two weeks for the time taken to issue the decision notice.

Before receiving a decision, there are two other matters to consider. Firstly, one may end up in long negotiations with the authority over the terms of any Section 106 Agreement. Secondly, the authority may ask to extend the period for their deliberations or extend consultation. In certain circumstances you may consider a dual application forcing one down the Appeal route, while the other application makes its normal progress.

If now you receive a refusal or consent with unacceptable conditions, you will have six months in which to appeal. Do you wish to appeal? What are the costs, likely outcome and potential delays? Appeals are where the most unpredictable time delays occur. An appeal may be conducted by written representations or a full Inquiry with an inspector appointed by the Secretary of State, and the appeal may either be delegated to the inspector to decide, or referred to the Secretary of State. The larger the construction project, the more likely it will be that a public inquiry will be needed, and in the case of any major controversial project, the case is likely to be called in by the Secretary of State for his own decision.

The pre-planning framework, consultation and identification of problem areas really comes into its own during an appeal stage. Addressing issues and preparing evidence, as well as reducing the number of objectors in the process, cuts down costs and delays.

From the lodging of an Appeal to the receipt of the Secretary of State's decision can sometimes take two years or even longer. Once the public inquiry has closed, or exchanges of representations completed, there may be nothing more to do than wait for the decision by the Inspector or the Secretary of State. But if substantial delay ensues, it is expedient to write to the Inspector or the Department to press for a decision. However, proceedings are not necessarily at an end at that stage. If the Secretary of State decides to make a ruling contrary in whole or part to the recommendation of an inspector reporting to him, or he considers that new facts and issues have come to light, he may launch a fresh round of consultations with the parties, and may even deem it necessary to re-open the public inquiry. It is therefore important not to disband one's support team in advance of getting a final decision.

Indeed the whole key to a smooth journey through the planning process is preparation, consultation and 'good planning'. Identify the issues in good time and address them. One should then be in a better position to answer the original question 'when will we be able to start work?'

4 PROJECT FINANCE

MALCOLM CLARK & MARK TINSLEY
County NatWest Limited

County NatWest recently arranged and co-ordinated the start-up financing for Thames Estuary Terminals, a container port project at the Isle of Grain.

This chapter on project finance is intended to act as a guide to the main issues associated with funding a major project and the practical implications for the project manager.

The term project finance generally applies to a specific project as opposed to, say, a company's continuing capital investment programme. The financing of a project can be described as 'the arrangement of adequate funds over time to finance the development and operation of a clearly defined project'. In practice the arrangements for a typical project financing will be required to fund the capital expenditure and working capital during the initial phases and to be repaid once in operation from income generated. Examples of projects frequently financed in this fashion are oil production facilities, mines, process plants and toll bridges.

Governments are frequent major sponsors of projects in both the developing and developed countries. Although the UK Government is seeking, in highways and other fields, to explore the possibilities for more funding of government projects from private finance, the great majority of government-sponsored projects in this country, as indeed abroad, tend to be funded out of general government revenue or borrowings, rather than with finance specific to the project. Similarly, many projects are undertaken by major companies whose financial strength enables them to finance projects from general corporate funding.

However, this chapter will for the main part deal with project financing where financiers look to the project alone for repayment.

PROJECT RISK

The perceived risk associated with a project will be critical to the sponsors' success in attracting project finance. Banks, usually providers of the majority of the financing, will assess the merits of the project in terms of the certainty of repayment.

A new project has no historical information on which to base judgements about its likely future financial performance. As a result, financiers will judge the merits of the project after undertaking a thorough analysis of the business plan and assessment of the track records of the management and contractors.

As a project develops from the engineering and construction phase, through the start-up phase, to the operational phase and beyond there will be a range of risks to which its financiers may be exposed.

These will include:

Completion risk: that the construction is not completed on time or on budget, for technical or management reasons.

Operating risk: that the project does not perform operationally as planned, for technical or management reasons, resulting in higher costs or lower output.

Market risk: that demand for the output is less than anticipated.

Supply risk: that the necessary raw materials (in particular, reserves in a natural resource project) are inadequate or too costly to recover.

Regulatory risk: that the regulatory environment changes which results in higher costs or other difficulties.

The project manager will be particularly concerned with completion risk which includes design risk, delays in construction, and cost overruns.

Providers of project finance will normally prepare financial projections based on the business plan. They will generally be concerned about the robustness of the projected cash flows in the face of these risks. As part of their analysis of the project, potential lenders will:

- ▲ assess the validity of the underlying assumptions made about these risks;
- ▲ test the sensitivity of the projected cash flows to these technical and economic assumptions, in particular the assumptions made about market risks such as future sales, prices and competition.

As part of their risk assessment exercise, financiers will usually require an independent appraisal of the risks to which the project is likely to be exposed from specialists in relevant fields such as engineering firms, economic or market consultants and technical experts. Certain risks can be mitigated by contractual arrangements such as fixed construction contracts (completion risk), or long term 'take or pay' contracts for the project's output (market risk).

In practice as certain financial institutions gain experience and develop a deeper understanding of the 'risks' associated with certain types of projects they may become more willing to assume certain risks previously unacceptable to them. The experience of banks lending to North Sea oil field development illustrates this well. In the early field developments banks would only take reservoir risk — ie that the oil was there. In later developments more technical and economic risks were taken.

The participating banks in the £5 billion Eurotunnel loan facility, the largest ever project financing, were invited to take the risk associated with not only the construction (completion risk), but with the completed tunnel's ability to attract the forecast levels of traffic (market risk).

RAISING THE FINANCE

Requirements

In most projects the main elements of the funding requirement will be:

- ▲ construction finance, to pay for capital expenditure;
- ▲ contingency finance, to pay for cost overruns and delays;
- ▲ working capital finance, to invest in stock and fund operating costs etc during the early phase of operations.

Although the structure and form of the finance will, to a large extent, be governed by the nature of the project, it remains likely that the finance will impose certain constraints on the design and scheduling of the project. It is therefore important that the project managers are aware of the nature of these constraints early on in the project design and planning processes. To this end, it is important that the project managers involve the financiers as early as possible in the development stages of the project.

Financing Structure

The most common project financing structure is 'limited recourse financing' where the sponsor's financial commitment is limited to an equity investment in the project. The remaining funding requirement is raised by way of debt which will be repaid only out of future project cash flows and directly secured on the assets of the project with no further recourse to the sponsor.

Frequently, lenders may require contingency finance for cost overruns or delays to be underwritten by other investors. This may take the form of a guarantee given by the sponsor or the sponsor's bankers to the providers of construction finance that interest payments will be met, and/or by the contractors to the financiers that the contract price will be fixed.

The borrowing cost of limited recourse financing arrangements will typically be higher than a major company's general corporate borrowings reflecting the additional risk that the lender is required to take.

For large projects which require an equity stake greater than an individual sponsor might be willing to invest, or where the risk exposure is too high for one sponsor, a joint venture partnership or consortium of sponsors is frequently established.

For example, natural resource development projects such as oil production or gold mines are frequently undertaken as a joint venture.

Sources of Finance

Funds are generally available from a range of lenders or investors, each of which has an interest in providing finance on different terms.

Equity investors, usually the project sponsors and their partners, but also sometimes including institutional investors, generally expect a much higher return on their investment than lending banks.

The banks, in particular specialist project finance banks, have traditionally been providers of term loans to projects with a relatively low return over their cost of funds. The flexibility of bank debt makes it particularly suitable for the contingency and working capital financing elements.

Managing Financial Risk

Apart from the more straightforward forms of bank credit facilities there are a number of other financing techniques which may be used to control interest rate risk, currency risk and improve the project's cash flow profile.

Of particular interest in recent years have been a number of new developments in financial markets which permit more sophisticated management of interest rate risk. Medium-term fixed interest loans have been available for some time in banking markets, but have not proved sufficiently flexible for many projects.

Interest rate swaps, where the obligation to pay floating rate interest on a loan is exchanged with a fixed rate borrower for his obligation to pay fixed rates, have become a common and more flexible option for hedging interest rate risk. Swap rates reflect the financial markets' perception of future interest rates.

Interest rate caps can be purchased for a cash premium, which will effectively set a ceiling to a floating interest rate for a certain period. In practice, the seller of the cap will be a financial institution which is willing to take a risk on future interest rates, and will pay the buyer the difference between the cap rate and the floating rate obligation if rates rise above the trigger level. Cap prices can vary dramatically, depending on both the gap between current rates and the chosen gap rate, and the level of uncertainty in the money markets. This is probably the most flexible technique, as the cap can always be resold to another buyer if it is no longer wanted.

Equipment and capital goods leasing can be effective in both reducing financing costs, as leasing companies can make better use of certain tax advantages, and improving the project's cash flow in the critical early years of the project by extending the terms of repayment.

PROJECT DEVELOPMENT

As the project develops from an idea to realisation, different project financing issues become important at particular stages of its development.

Pre-Implementation

The financing process will be well developed even before the project itself has left the drawing board. The project sponsors will generally have undertaken a thorough economic and financial evaluation of the project, often incorporating this in a business plan with which to approach the banks and possible co-investors. At this stage, detailed costings will have been done, future cash flows assessed and the major elements of the financing requirements will have been identified: namely, capital expenditure, contingency and working capital.

In addition, interest payments may often be required to be 'rolled up' until the project becomes cash positive and the repayment schedule will be designed to reflect the project's future cash flows.

At this stage, the project sponsors will approach a number of banks or investors and seek terms for the finance. After the financial institutions have assessed the risks, offers will be negotiated and agreed. This will often result in a financing structure significantly amended from the sponsor's original proposals.

Project Construction

The actual availability of the funding will usually be subject to a number of conditions such as satisfactory independent consultants' reports, prior subscription of the equity, agreement of all important contracts including that of the project manager, and the recruitment of the management team.

In addition, each drawing of funds will usually require evidence that expenditure is due in the form of an architect's payment certificate or certified cost certificate signed-off either by the project manager or an independent adviser to the providers of finance.

Ongoing Controls

Project financing will be subject to certain ongoing controls imposed by the banks. Often in the form of financial performance ratios, these provide an early warning system to the banks that the project is not performing according to plan. Project-discounted cash flow to loan value ratios, the ability to pay interest and repayments out of cash flow, capital expenditure limits, and schedules by which progress or performance can be measured, are typical examples of such controls. Independent consultants may advise the financiers as to whether controls are being adhered to and targets are being met.

PROJECT MANAGER'S ROLE

As mentioned above, the project manager, in addition to his principal responsibilities would typically be involved with regular monitoring and evaluation of every aspect of construction and communicating this to the financiers or their advisers.

The key issues for providers of finance are:

- ▲ planning and design of the project;
- ▲ scheduling and delay management;
- ▲ accounting for expenditure and progress;
- ▲ budgeting for contingencies;
- ▲ managing contractors;
- ▲ information control and reporting.

The project's financial structure is likely to have a distinct impact on the nature of the project manager's reporting role. Where the project is fully funded by the sponsor, his communicating and reporting responsibilities are likely to be less formalised and extensive.

The project manager's role in budget management is also affected by the form of contract involved, which is in turn affected by the nature of the financing arrangement. Where the construction contracts are fixed price rather than cost plus, there is likely to be less need for the project manager to focus on budgets, and more on the quality of work, although when contingencies arise during construction, the fixed price will often come under pressure.

In broad terms the more project risk that the financiers and sponsors have assumed, the more they will expect regular detailed flow of information about expenditure and progress.

The project manager, in such a central position during a project's construction phase, bears, in the view of the financiers, a substantial responsibility for the management of all aspects of completion risk.

5 CONTRACT FORMS AND STRATEGY

BOB MOYSES
Franklin and Andrews

INTRODUCTION

Forms of contract are required to formalise the complex arrangements and relationships that are likely to be encountered in the construction of all major projects.

A form of contract is the document that details the contractual terms to be applied between the contracting parties. It identifies the various documents which together represent the legally binding contract. These will comprise some or all of the following:

- ▲ the agreement
- ▲ the conditions
- ▲ the tender
- ▲ the pricing provisions
- ▲ the specification
- ▲ the drawings (if applicable).

In simple terms, the form of contract specifies the responsibilities, liabilities, method of payment and apportionment of risk between the parties in achieving the desired objectives relating to quality, time and cost.

There are various aspects of a project for which forms of contract may be applicable and these may be dealt with individually or as a whole. These fall into the following broad elements:

- ▲ design/engineering
- ▲ material procurement
- ▲ construction
- ▲ commissioning/handover.

The contracting philosophy chosen by the client will determine how these areas of activity are to be handled and to a large degree will depend on the level of involvement and control he wishes to have in the execution of the contracts. The client may also decide in particular circumstances to be the managing contractor.

In deciding his contracting philosophy the client will have to consider the relative importance of factors such as:

- ▲ timing
- ▲ the ability to make variations
- ▲ the level of quality
- ▲ the certainty of price
- ▲ the apportionment of risk
- ▲ the complexity of the project
- ▲ the management style required.

The client will therefore need to make a decision as to which form of contract should be adopted in order to execute the required contracts. This decision may be made on the basis of his own experience on previous projects or may be on the advice of appointed consultants. In either case the factors affecting the choice of form of contract, as detailed later in this chapter will need to be considered.

Contracts will then need to be established between some or all of the following parties:

Managing Contractor(s) (including the client as managing contractor):	To execute or supervise all elements of the project.
Contractors:	To execute some or all of the elements of the project under contract to either the client or managing contractor.
Subcontractors:	To execute works under contract to the managing contractor or contractors.
Consultants:	To undertake specialist activities such as design and construction supervision, under contract to the managing contractor or client.

Certain forms of contract are published by various recognised bodies and are available 'off the shelf'. These are known as Standard Forms. The alternative to using Standard Forms is the production of non-standard forms to suit a particular requirement.

STANDARD FORMS OF CONTRACT

The advantage of using Standard Forms is that they have generally been tried and tested over many years with recognised legal precedents having been established. The forms are subject to constant review and amendments are made to incorporate any recent developments.

It is therefore important to ensure that, whichever form is chosen, the latest revision is considered.

If it is decided to modify a Standard Form, great care must be taken to ensure that the integrity of the document is maintained, taking account of the cross-referencing and inter-dependency of the clauses.

Care will also need to be taken to ensure that any modifications to standard forms, or clauses produced for non-standard forms, do not give an undue bias to either party, as in the long-term this is likely to be counter-productive.

When choosing a Standard Form of contract, the client needs to be aware of the different contract philosophies embodied in the documents, particularly with regard to the use of third-party supervision. For example, the Joint Contracts Tribunal (JCT) and Institution of Civil Engineers (ICE) forms allow for a supervisory role to be carried out by a third party (architect or engineer) and the person so appointed is meant to execute his duties within the terms of the contract with impartiality and integrity towards both parties to the contract. This role has been described as that of a 'quasi arbitrator'. Whether such intent is achievable in practical terms has been a matter of debate over many years.

Other Standard Forms (eg Institution of Chemical Engineers) assume a simple client/contractor relationship.

The nature of the contracts to be placed will also affect the choice of Standard Form (eg for the supply of equipment, the supply of a service, the execution of works including the supply of materials and labour).

Standard contracts are aimed at particular sections of the construction industry and are structured accordingly. In broad terms they can be classified under four headings:

- ▲ Building
- ▲ Civil Engineering
- ▲ Chemical and Heavy Engineering (including specialist mechanical and electrical services)
- ▲ Government.

The main factors to be considered in the choice of the Form of Contract can therefore be summarised as:

- ▲ Suitability for the particular project
- ▲ The purpose of the contract
- ▲ The proven track record of the document
- ▲ Familiarity of use within the industry.

There are numerous Standard Forms of Contract available, some of which have options available depending on the size of the contract or type of pricing provisions, together with associated standard documents for use by sub-contractors.

The following table outlines the forms that are most commonly adopted on the majority of projects and provides in broad terms the type of project to which they would be best suited.

TITLE OF DOCUMENT	TYPICAL USAGE
JCT Contracts — These are contracts published by the 'Joint Contracts Tribunal for the Standard Form of Building Contract'.	Building Contracts of all sizes and types allowing for various forms of pricing provisions.
There are numerous versions which cover management contracts as well as contracts with contractors and subcontractors.	Also applicable for management and design and build contracts.
ICE Contract — The full title is the 'Conditions of Contract and Forms of Tender, Agreement and Bond for use in connection with works of Civil Engineering Construction, issued by the Institution of Civil Engineers, the Association of Consulting Engineers and the Federation of Civil Engineering Contractors'.	Civil engineering contracts of all types. In amended form, may also be used on multi-disciplined projects for all engineering contracts.
FIDIC — This is short for 'Federation Internationale d'Ingenieurs Conseils'. In essence the contract is the overseas version of the Civil Engineering Contract.	As ICE for use overseas.
GC Works 1 — This is a contract published by Her Majesty's Stationery Office, the full title being 'General Conditions of Government Contracts for Building and Civil Engineering Works'.	Large building and civil engineering contracts undertaken by Government departments and certain major authorities.
Institution Chem.E — This is a contract published by The Institution of Chemical Engineers, the full title being 'Model Form of Conditions of Contract for Process Plants'. There are two versions, one for lump sum contracts, the other for reimbursable contracts.	Multi-disciplined projects covering all engineering contracts. Particularly suited to the construction of process plants.
ACA Form — This is a form of contract published by the Association of Consultant Architects. It is a simplified building contract.	Building contracts.

TITLE OF DOCUMENT	TYPICAL USAGE
Model Form — This is a term used for more than one type of contract but is generally understood to mean those contracts issued jointly under the recommendation of the Institution of Mechanical Engineers, The Institution of Electrical Engineers and the Association of Consulting Engineers. There are a number of variations (Model A, Model B etc) each serving a different purpose.	The design and construction of specific engineering works.

Variables to Standard Forms

Within the framework of a Standard Form, there will be a number of items that will require specific input to suit a particular contract, such as:

- ▲ times and stages of completion
- ▲ taking over procedures
- ▲ performance testing
- ▲ insurances
- ▲ retention
- ▲ security bonds and guarantees
- ▲ liquidated damages.

Other items that may not be covered in the Standard Form in detail must be included elsewhere, usually in the pricing provision, eg The Bill of Quantities. These items include:

- ▲ description of the works
- ▲ description of site
- ▲ drawings used for the tender
- ▲ tender information
- ▲ services and facilities
- ▲ working with other contractors
- ▲ temporary works
- ▲ recording of information
- ▲ protection

- medical and first aid requirements
- security.

These items will also have to be catered for in the production of non-standard forms in conjunction with the clauses listed in the next section.

NON-STANDARD FORMS OF CONTRACT

There are many non-standard forms of contract which have been produced by various companies in their capacity as either clients or managing contractors and these should be considered on their merits in line with the requirements of the individual project.

The advantage of using or producing a non-standard form of contract is that it can be tailored to suit the specific needs of the project. This factor can be of great importance in specialised types of development using advanced forms of technology.

When drafting non-standard forms there are certain clauses that should be considered as fundamental and these can be listed in broad terms as follows (not in order of priority):

- definition of terms
- contract documents
- general obligations
- assignment and subcontracting
- programme of work
- delays and extensions of time
- damages for delay
- maintenance and defects
- variations
- inspection
- working conditions/safety requirements
- insurances
- taking over
- payment
- termination
- arbitration/settlement of disputes.

Having drafted a non-standard form of contract which is considered to satisfy all the specific requirements of a particular project, it is recommended that independent legal advice be considered to check on the contractual integrity of the document prior to its implementation.

PRICING PROVISIONS AND STRATEGY

One of the most important parts of the contracting philosophy developed for a project is the strategy to be adopted with regard to the pricing provisions. The chosen pricing provision will be a factor that will then be taken into consideration in the selection of the type of form of contract to be adopted.

Relevant to choice of any of the pricing provisions to be described is whether the work is to be undertaken by a single multi-disciplined contractor or by a series of contractors responsible for various aspects of the project. This will affect how, in the early stages of project planning the outline work scope is developed and divided into individual contract packages for which tenders are to be obtained.

The decision on the type of pricing provision to be adopted needs careful consideration as it will have an effect on all aspects of the commercial management of the project throughout its life and the resultant price paid. The factors that will affect this decision will include such items as the degree of engineering or design definition available at the time of production of contract documents, the nature and size of the project and to some degree the general economic circumstances pertaining at the time.

Other points which may affect the decision to a lesser degree would be the time period available to produce tender documentation and the time period planned to execute the work.

Having considered the above points, the pricing provisions from which a choice is made are:

- ▲ lump sum or 'turnkey'
- ▲ target lump sum
- ▲ bill of quantities
- ▲ schedule of rates
- ▲ reimbursable or a combination of certain of these options.

Factors affecting the choice of pricing provision are many and can be summarised for each type of pricing provision as follows:

If the work scope and specification is defined in detail or the project is to construct a standard type of facility, a fixed lump sum or 'turnkey' type of contract may be favoured. This has the advantage to the client that he knows with a fair degree of certainty his financial exposure having the management and risk element in the work scope passed on to the contractor. By doing this, however, the client must recognise that he will not be able to make changes to his requirements during the execution of the works or impose his views (other than those agreed at contract award) on how the project is controlled, eg if the conditions of contract have been written to allow the ordering of variations by the client then alterations to requirements can be made within the parameters of that variation clause. However, this will to some degree detract from the main purpose of a fixed lump sum approach due to the inevitable departure from budget and possible effect on schedule.

Before embarking on a turnkey strategy the client must satisfy himself that the contractors tendering on this basis have the capability to complete the works to the required specification, as failure to perform will affect the whole project. The client has passed on the risk, but he has also effectively 'put all his eggs in one basket'. The target lump sum approach may be set up with a single contractor for the whole of the works, or with several contractors being given specific areas of work.

The contractor prepares an estimate which is related to a defined scope of work. This estimate is reviewed by the client and following negotiation is agreed by the parties. If the price at completion is greater or less than the pre-agreed estimate, the difference, whether 'profit' or 'loss', is shared in predetermined proportions by the parties. The target method of payment gives the contractor an incentive to operate efficiently, as he has access to cost savings in 'profits' but reduces his liability with regard to cost overruns, ie 'losses', as the client is sharing some of the risk. The client gains advantage as the contractor is encouraged by financial incentive to share the client's goals of earlier completion at least cost.

This type of payment has most advantage where close links between client and contractor are of particular importance, eg 'fast-track' construction where design and construction proceed alongside each other, or where novel methods of construction are to be used, which if successful have considerable commercial benefit to the client. The use of bills of quantities was initially confined to building and civil engineering projects, but since the early 1950s has become more widely used for all engineering disciplines.

This approach involves the measurement of construction activities in accordance with standard or non-standard methods of measurement so that when the bills are priced their total forms a sum on which the contract is based.

The bill of quantities main advantage is that it is a flexible document for accommodating change to either workscope or specification. It enables remeasurement of specific quantities or revision of unit rates on a predetermined basis.

The bill of quantities gives a consistent basis for tendering, enables evaluation of partly completed works for interim payments and gives a basis for the valuation of variations. It can be used as the prime element in post-contract cost monitoring and cost control systems.

If the pricing of the bill of quantities and the resultant sum is to be realistic, the design/engineering information must be complete or reasonably well advanced at the time of document production.

The contracting plan will have to accommodate the time necessary for the quantity surveyors to prepare the bill(s) of quantities which may put back the date for receiving tenders, but this time will normally be countered by the reduced period which may be required for tender assessment and tender clarification discussions.

In cases where the design/engineering information is not well advanced, approximate bills of quantities may be compiled, based on historical data from similar projects. This will enable those tendering to have some idea as to the overall scope of work. Where the design/engineering information is poor or where the activities are known but the scope cannot be foreseen because of the nature of the work, eg modifications to existing units, it is more advisable to adopt the use of a schedule of rates.

Like the bill of quantities, the schedule of rates gives the contractor an incentive to be efficient as he will be paid for completion of work on a unit rate basis. It is more likely, however, that the rates will be higher than those submitted in a bill of quantities as the contractor will include a higher contingency factor due to the unknown scope of work. In certain circumstances, a contractor will need to be appointed to execute work for which there is no definition available either with regard to scope or type of work at the time of obtaining tenders. An example of such a situation is the carrying out of shutdown work on chemical or oil installations where the extent of the work is unknown until the plant is shut down and partly dismantled.

In these cases, there is little option but to carry out the work on a reimbursable basis. A clear definition of what will be included in the reimbursement will need to be established in order to obtain competitive tenders.

This form of approach gives the client direct control over the activities of the contractor, and hence the level of expenditure undertaken. The contractors should be able to tender competitive rates; they will not need to build in contingency sums as the risk element is carried in total by the client.

It will not always be possible to make decisions as to which type of pricing provision should be adopted on clear cut situations as outlined here, and an element of judgement will always be required.

This is because of the complex nature of the design, procurement, construction and handover aspects that can be encountered in the control of major projects which to some degree are all unique in their requirements.

As mentioned earlier, it may well be the case on major projects that a combination of various options will be most suitable.

The subject of contract forms and contracting strategy is a broad and complex subject. The points raised in this chapter do not constitute an exhaustive list of contract types or of the features which can be built into contracts; they will, however, given an indication or guideline into the decision making process.

The following publications are recommended for further reference:

TITLE	**PUBLISHER**
Thinking About Building	NEDC Building Economic Development Committee
Practice Note 20 Deciding on the appropriate form of J.C.T. Main Contract	Joint Contracts Tribunal for the Standard Form of Building Contract
Guide and Flowchart for selecting the appropriate J.C.T. Form of Contract	B.E.C.
Tenders and Contracts for Building	The Aqua Group

6 PROJECT CONTROL AND ORGANISATION

ALAN WILLS
Fluor Daniel Limited

PRINCIPLES OF PROJECT ORGANISATION

A project is a unique event and therefore a project organisation must be conceived and implemented which will bring together a team of personnel from various disciplines, some of whom may be from separate organisations. It is important that the organisation reflects the contractual arrangements, the scope of work, and the required communications, and also interfaces between the owner, contractor(s), and any third parties.

As initiator and owner of the project, the client must undertake a series of fundamental steps. These include:

- ▲ undertaking the necessary feasibility studies and budget estimates (using consultants as necessary) to enable him to make the decision to proceed.
- ▲ ensuring finance is available and to pay the bills.
- ▲ making the site available.
- ▲ defining the project. This could be a functional specification, or it could go down to such detailed levels as size of rooms and main parameters of services. The important point is that *all* the owner's requirements are specified such that designers can proceed with the confidence that there will not be substantial changes later.
- ▲ preparing a preliminary contract strategy. This will be finalised after appointment of the project manager.
- ▲ appointing the project manager (PM). The PM's terms of reference will depend on the preliminary contract strategy. In some larger organisations, the owner may do his own project management; in others it is provided by a consultant/or contractor.
- ▲ determining the levels of delegation to the PM. What has to come back to the owner for approval or decision?
- ▲ agreeing the final contract strategy with the PM.
- ▲ approving the appointment of other participants, arranged and managed by the PM.

The client must also provide a support structure for the project manager, including monthly cost and progress reports, change control, and decisions above the delegated limits. He should prepare for the utilisation of the completed project, and participate in commissioning and handover.

The project manager is the key individual in all of these matters. He must be fully aware of the client's project goals, parameters, and criteria. The project manager should be given the authority and responsibility for planning and controlling the project work as well as being held accountable for overall results. Authority and responsibility can be delegated by the project manager to lower levels in the organisation. However, accountability for the project must lie solely with the project manager; it cannot be delegated.

Successful project management to realise goal achievement during one-time projects requires integrating cross-functional groups within an organisation and working with external groups to achieve stated objectives. It is quite common on major projects (particularly involving joint ventures or similar arrangements) to set up an executive steering committee to oversee the performance of a project. The steering committee would normally consist of senior management representatives of the client, managing contractor(s) and any joint venture partners.

CRITICAL ISSUES

Critical issues are those that could jeopardise the success of the project if not satisfactorily handled. Often these are identified in the Owners' Invitation to Bid. Others may be identified during the contractors' planning, normally prior to contract. They will vary from project to project but the following are typical issues: authority approvals of design, construction and environmental issues, funding, grants, locality, counter-trade requirements, site constraints, material logistics, modular construction, safety issues, policies and attitudes of government departments and local authorities, public relations, manpower, industrial relations, weight control and climatic constraints.

Some of these critical issues can be handled by special attention to normal project controls. Others however may either be so important or sufficiently unusual that the project organisation must be adapted to ensure that they receive adequate or individual attention. For example, counter-trade may necessitate an assignment of a specialist in that area and modularisation may warrant the appointment of a manager of modular design/construction.

PROJECT CONTROLS

An important necessity for achieving the project objective is developing a timely realistic, and useful master project plan that incorporates the project's scope, cost, dependent activities, resource requirements, and resource availability.

Project control provides the information that is necessary for reporting on the project status both internally and to the owner. The principle project controls are estimating, scheduling, cost and change control.

Control is necessary at all interfaces between the project team and the owner, and between the project and its external environment, eg battery limit connections.

The following criteria are most commonly used to measure a project's success:

- ▲ Client satisfaction
- ▲ Performance to specifications in terms of output and quality
- ▲ Budget performance
- ▲ Schedule performance
- ▲ Contractor and project team satisfaction.

Characteristics of successful projects are associated with the good management of planning, goal commitment, team motivation, technical competence, scope and work definition and project control systems. Poor project execution is associated with an unrealistic project plan, client/management changes, insufficient front end planning, and underestimating the project scope.

Planning is the identification of what needs to be done to meet the project objectives, while monitoring of progress identifies what has actually been done, and control compares the actual performance to the plan and thereby enables identification of variances. This allows project managers to evaluate selectively (using management by exception) the impact on project goals. Feedback gives management the opportunity to change either the plan or the execution of the plan. Constant change is inevitable and continual iteration through this process is always necessary.

MASTER PROJECT PLAN

The development of a Master Project Plan normally follows the sequence of basic engineering, detailed design, procurement and construction. On fast track projects these phases become overlapped such that the next phase commences significantly before the completion of the preceding phase.

MASTER PROJECT PLAN DEVELOPMENT STEPS

The client's Invitation to Bid which normally begins the project work should include expected performance requirements, cost limitations, schedule requirements, and contract arrangements. Additionally there should be a statement of objectives to define adequately the overall project goals including overall project specifications and necessary client requirements so that the participants fully understand the work to be performed. The Statement of Objectives should include the deliverable end items, quantifiable objectives, measurement of objectives, the scope of the project cost target, preliminary start and end dates, legal obligations, financial obligations, and any predetermined supplier relationships. These objectives should be reviewed between the client's project manager and the contractor so that there is a full understanding of the scope of the project.

The management level schedule shows planned dates for major activities. It consists typically of five to 30 activities, depending on project size, which identify the project start and end dates and other major milestones. It should initially be used at the management level to plan personnel requirements in support of the project.

An order of magnitude estimate (also called the conceptual, factored, or feasibility estimate) is based on little or no detailed design or planning of the project. This estimate may be based on a similar project but adjusted for present day prices and the productivity in the area where the project will be performed.

The Statement of Objectives, Management Level Schedule and Order of Magnitude Estimate are used to obtain client and management approvals.

Work Breakdown Structure

A basic management technique is to break a project into manageable pieces in the Work Breakdown Structure. A WBS is a hierarchically structured listing of project activities (ie physical items, procedures, services etc) broken down into work packages. The WBS is an important responsibility of the project manager because it provides a common framework. The total project may then be defined at a detailed level and a common scope/cost/schedule framework may be used for communication, assigning responsibility, work authorisation, planning, monitoring and control.

Work package budgets should be identified in hours, material usage, etc and must be measurable.

Once the WBS has been finalised, the Code of Accounts, a numbering system for cost monitoring, control and forecasting, may be developed and based on the Work Breakdown Structure.

The number of work breakdown structure levels and the scope of each work package should be based on the project's size. In general the greater the project complexity and technical requirements, project cost and time span, the greater the number of work breakdown structure levels and work packages. Having the appropriate number of Work Packages is crucial to implement the work breakdown structure effectively.

Once approval is obtained for the Statement of Objectives, Management Level Schedule and the Order of Magnitude Estimate, the design/planning during the initial stage of the project may now begin.

Conceptual Design and Planning

Initial design and planning requires sufficient effort to define the project adequately. Examples of the type of documents normally produced during conceptual design for an engineering/construction project may include flow diagrams, equipment listings, a detailed description of work, and layout drawings.

Increasingly, the computer and particularly computer aided design (CAD), and computer aided engineering (CAE), is helping reduce the overall technical manhours and increase the access to conceptual design information once it is developed.

Budget Estimate

The Budget Estimate (also called appropriation, control or design estimate) is based on quantitative information using bid and unit rate prices for material, equipment, tools and the cost of manpower requirements. Due to the uncertainty of project work, the total project budget estimate may include provisions for potential scope changes, escalation, currency fluctuations, project manager's and corporate management's contingency.

A budget estimate may be developed based on a definition of the project's system and structures, flow sheets, layouts, equipment details, major equipment and material quotation information from vendors, special site considerations, availability of management and labour, current trends in material and labour escalation, and regulatory requirements. Depending on the industry and type of contract, the budget estimate should be within −10 per cent to +25 per cent of the actual final cost of the project.

Network Diagram

A Network Diagram identifies what tasks must be performed and their logical relationship and interdependencies. The level of activity detail when preparing the network diagram during this planning phase should follow the work breakdown structure at a summary level (ie more activities than the management level schedule detail), but fewer activities than that required to reflect adequately the project work scope at the work package level.

The goal should be to produce a network that may be used to plan the work of major functional groups on the project to a better degree. At this stage of the project, getting into too much activity detail is not appropriate because too many assumptions are often required. In addition, too much activity detail often detracts from project management's vision of the whole project.

The network diagram is very important since this document is central to the Master Project Plan. Activity relationships normally fall under three headings: normal, preferential, and external. Normal precedence relationships are hard and fast and normally nothing may be done to alter this sequence of work. The preferential activity relationships often represents the project manager's choice and may reflect a strategic, conservative, or even risky approach to project execution. External activity relationships are used to identify equipment or resource constraints (eg only one heavy lift crane, or labour saturation levels).

Usual resource requirements and availability should be used when estimating 'normal' activity durations. Once the network diagram has been completed, a forward and backward pass may be performed, possibly using a computer software package, to identify the critical path through the project. Management by exception of the identified critical path activities is the key to on-time project execution.

Resource Allocation

Resource Allocation requires loading individual resources; manhours by craft, number and type of equipment, and material quantities onto network activities. These resources may then be converted to cost figures using estimated labour rates, purchase costs for equipment and material, and unit rates. Completely defined resource loading is a major undertaking which is usually not required or even feasible within the limited time available. The approach should be to identify a maximum of four of the most potentially critical resources, eg labour manhours, tools, equipment, or material quantities to allocate to activities.

Engineering construction projects are normally constrained by labour availability and/or saturation, major equipment procurement and installation followed by critical materials. Once selected resource loading has been accomplished, time and resource availability through time have to be determined.

Two major scheduling approaches based on activity precedence and resource constraints are available. The first is time-constrained (also called time-limited) scheduling where the end date is based on the forward pass (ie earliest date possible) or an externally imposed date. The resulting resource load histograms are then used to identify any resource requirements which exceed availability. The second is resource-constrained (also called resource limited) scheduling where all resource availabilities are fixed and may not be exceeded for any reason. Activity start and finish dates are moved to conform to resource use and availability.

In order to minimise the expected time of completion dates, there is often a need for a time/resource trade-off. However, the relationship of time to resource trade-off *should not be assumed to be linear* (eg twice the resources, half the time). When other factors interfere, such as space limitations which force overcrowding leading to reduced productivity, excessive overtime, or increased indirect costs, resources which have been added may actually increase activity duration times.

The Master Project Plan may now be developed based on the network diagram, an expected completion date, and critical resource activity loading and constraints.

Master Project Schedule

After accounting for networking and resource constraint priorities, some activities typically remain floating which represent the 'remaining flexibility' in the project plan. Additional, but normally lower ranking project priorities may now be included such as the timing and size of construction contracts. The resolution of any conflicts remains the project manager's responsibility in selecting activity start and end dates and producing the approved finalised Master Project Schedule. Following final resource allocation and activity scheduling, the budget estimate may need some revision. An approved finalised Master Project Plan may now be used as the base project plan. An easy to update plan that is credible and that can adequately represent a rapidly changing project will give the project manager confidence to act when there is a variance from the plan.

Project Control Systems

Project progress should be regularly monitored and reported at the work package level which should have a specific limiting schedule start and completion date.

It is important to distinguish between physical progress and manhour expenditure in engineering and procurement work. Physical progress is best measured by assigning percentages to key milestones in the work at the control level and to measure actual percentage complete by reference to those milestones.

For example, a drawing may be considered 10 per cent complete when started, 50 per cent complete when first drafted, 70 per cent complete when first issued for comment, and 90 per cent complete when issued for construction, leaving 10 per cent for subsequent revisions and site changes. By this means it is possible to calculate earned manhours as a measured percentage of the budgeted hours.

The ratio of earned hours to actual hours spent will give a measure of productivity. If productivity is less than 90 per cent at any given time it will indicate either an inadequate budget or inefficient working. In either case, additional manpower may be necessary to recover the position.

Similarly for site work, physical progress must be the basis of progress measurement, not manhours expended. Site work is normally measured in terms of cubic metres of concrete, tonnes of steel, numbers of items of equipment and metres of pipe and cables installed.

In order to ensure that adequate resources are available, it is essential to re-examine the remaining work to complete, and to re-forecast the hours required for that work. It is important to avoid underestimating the remaining work in order to compensate for low productivity on work completed. If productivity has been low on completed work, it is likely to continue to be low. This unpalatable fact, if ignored, will lead to substantial underestimate of manhours' expenditure and thus cost, as well as a failure to complete the work on time due to inadequate resources.

Accurate measurement of progress, productivity and forecasts to completion takes significant effort and therefore is usually carried out on a monthly basis. However, if necessary, it can be carried out fortnightly or even weekly. Where it is necessary to monitor particularly critical activities more frequently, a separate manual control system is often used with the normal computerised systems retained for monthly reporting. More frequent reporting is particularly necessary on activities that require a high manpower expenditure over a relatively short period, such as for the production of piping isometrics.

Recovery of Delays

It is much easier to avoid delay by pro-active management of early activities, than to try and recover delays later. Generally speaking every effort should be made to complete home office work on time, rather than to leave the recovery of delay to the field where recovery will be more difficult and significantly more expensive. The best way of avoiding delays is to establish a realistic project schedule, identify critical areas and establish a thorough system of measurement of progress, so that problems can be identified at the earliest possible opportunity.

If delays occur in non-critical activities it may be possible to accommodate them by the use of float. However, this should be avoided, particularly at the early stages of a project since float used then is not then available for subsequent activities. Delays may be recoverable by the increase of resources, whether of manpower or equipment. If this is not possible then it will be necessary to re-develop a plan which changes the logic of some subsequent activities.

Finally, it has to be recognised that some delays may be irrecoverable, in which case the important thing is to forecast accurately the consequences and work to minimise the effects.

COST CONTROL

Cost control covers the preparation of estimates, the monitoring of expenditure, the forecasting of costs and the issuance of regular reports.

Cost Control System

A project coding system must be established by the contractor. Depending on the type of contract this may be the owner's system, contractor's system or even a project specific system, incorporating where possible a cross-reference between the owner's system and contractor's system. However difficulties may be experienced where the breakdown of codes between the two systems is not the same.

Generally the system will segregate costs for various types of equipment items, bulk materials and construction cost by discipline. The project cost control system should interface with the accounts payable system and this requirement may suggest the cost code system to be used.

Costs incurred must be recorded promptly as committed (ie not when invoiced). Normally when a commitment is entered into, such as a purchase order for equipment, the entire value of the purchase order will be reported as committed. An exception may be made if a cancellation cost has been agreed, in which case the cancellation cost can be considered as the commitment for the duration of the cancellation cost agreement. Purchase orders should be structured to reflect the cost control budget and cost coding system. If equipment cost items are being controlled by item number then the purchase order must give total item cost and not grouped costs.

Extra work is a particularly important subject for cost control. It may be work that was allowed for in the budget, but not included in the contractor's scope, or it may be additional to the total project scope. It may have to be offset against the contingency or it may warrant a revision to the project budget. The important thing is to establish a cost control system based on a detailed budget so that extra work can be identified.

Sampling is a method of checking that the intent of the project execution plan and budget is being followed. It involves taking sample drawings or specifications, estimating the quantities and costs contained therein and comparing them with the budget quantities and costs. Sampling when carried out at the right time can show deviations from plan and budget in sufficient time to consider possible re-design in order to reduce quantities and/or costs.

The cost control system should allow for cost forecasting. Forecasting should be based on actual costs where known and the best possible estimate of costs yet to be committed. If existing commitments are showing an over-run, it may be appropriate to forecast an over-run on similar uncommitted costs. The forecast of uncommitted costs is the joint responsibility of the project cost engineer and the project manager. This is particularly important for subcontracts where large sums of money are involved. In such cases a regular review of the forecast subcontract cost should be carried out by the project manager, construction manager, contracts engineer and cost control engineer. The review should address the subcontract committed cost, and forecast the final costs by considering additional costs resulting from extra work, delay, rework etc.

On a major project, costs may be expended in a variety of currencies. Normally all costs should be converted to one currency at fixed project exchange rates for cost reporting purposes. However, separate reports may be required for forecasting and recording the effect of exchange rate variations.

Cost Reporting

On any major project a detailed cost report must be produced by the contractor. The level of detail will depend on the type of contract, but normally the contractor will report commitments and forecast against all project costs whether budgeted or not.

Except for lump sum projects the contractor will issue the owner with a detailed monthly cost report by cost code item and a summary level report within a monthly project report. In addition, the contractor may need to supply a forecast of cash requirements to meet committed expenditure.

The contractor may also prepare detailed cost reports for expenditure so that he can monitor his own costs. The contractor will also normally prepare an internal cash flow report covering his own contract, showing forecast and actual expenditure and revenue.

For overall control, it is also desirable to report on the earned value of work done on the project. The earned value represents the financial cost of work actually done, which is not necessarily the same as the costs actually paid. Typically for equipment, the earned value is much higher than any progress payments, and for subcontracts, payments are usually at least one month in arrears of the earned value. The earned value as a percentage of the forecast cost can be used for comparison with the percentage complete. If there is a substantial difference between the percentages of earned value and progress, it suggests that one of the figures is suspect.

Cost Forecasting

The cost report will routinely show actual commitments against the budget. In order to prepare an overall project forecast, it is necessary to predict costs of work not committed. The first significant consideration is quantities. If the quantities are greater than allowed for in the estimate, then a corresponding cost over-run should be forecast. In addition, unit costs should be monitored and, if budgeted unit costs are clearly inaccurate, then forecast costs should be adjusted. Trends on one purchase order or contract can be used to predict costs on another purchase order or contract.

Cost Control Summary

Cost control depends on a budget structured to the project execution plan, forecasts of quantities and rates. Cost reports should be issued monthly. Detailed reviews of subcontract cost forecast should be carried out.

CHANGE CONTROL

A change to a project is anything which changes the scope of work, project cost or schedule. Wherever possible, changes should be avoided because of their negative effect on both cost and schedule. Also, wherever possible, the project schedule should allow adequate time for project optimisation at an early stage so as to avoid rework later.

It is important that all members of the project task force fully understand the project scope, budget and schedule so that they can identify changes. Some changes are inevitable and they can best be accommodated by early recognition. Late changes have an increased impact on project cost and schedule.

Once changes have been identified they must be approved or rejected promptly. This approval may involve the owner, depending upon the contract terms. Generally no work should be carried out on changes until they have been approved. After approval, a specific plan should be prepared in order to implement the change with minimum impact on the project. The costs of changes may be reported separately or by revision to the overall project budget.

It is important to recognise that changes may have cost and schedule implications. These should be clearly identified prior to approval of the change. *Some proposed changes may be so significant that it is better to carry out the change after completion of the remainder of the project.*

MANAGEMENT OF INTERFACES

All projects have interfaces with adjacent plant, utilities, infrastructure etc. These are physical interfaces that must be considered during design. A specific procedure should be prepared to cover tie-ins. In addition there will be interfaces between the project task force, the client, the contractor and third parties.

These interfaces must be identified and formalised by way of written procedures and document distribution charts. These will cover preparation and issuance of documents, review and approval of drawings and documents.

CONCLUSION

Every project is different and therefore, to ensure success, each project must be organised, planned, controlled and reported on in detail. The success will largely depend on the effectiveness of management, planning and control carried out.

Good communication is vital to success, especially in view of the matrix management organisation.

7 QUALITY

BRYAN BURDEN
Shell UK Limited

There are two essential ingredients for the achievement of quality standards: people and systems.

Achievement through people requires complete commitment from the top by all parties concerned be they client, designer, contractor, sub-contractor or supplier. It requires the belief that quality is essential to the success of the project, not an optional extra; it requires strong leadership, adequate training and an attitude of mind based on pride in the job.

People cause quality problems by lack of an adequate brief, lack of a clearly defined and appropriate specification of what is required, lack of accurate, practical, buildable and timely drawings, lack of communication, and lack of care.

While formal quality systems in construction can never in themselves fully compensate for the shortcomings of people, they can assist them in raising their performance to the standard required.

Quality Assurance (QA) is a combination of two elements, quality plan and Quality Control (QC), to achieve an agreed quality objective. Quality Management (QM) requires the vital third element — improvement, ie learning from (expensively) acquired experience.

As many clients know to their cost, the existence of a legal contract is no guarantee that the client will be satisfied with the complete project. The client achieves small consolation from the eventual compensation through arbitration or the civil courts. Unlike when purchasing domestic goods the client cannot obtain a satisfactory replacement, and may be left with a patched-up original that results in high maintenance costs for many years to come.

QA is needed to demonstrate that adequate systems and procedures are in place to ensure effective control of the project. QM is needed to ensure continuous improvement in the standards of products and services.

QC is required to build quality into the design, planning and execution of all stages of the project to ensure safe design, safe construction and safe plant operation. If short-cuts are taken in the areas of QC and QA, disastrous consequences may result, eg lengthy delays and associated high costs due to late delivery of wrong materials, design errors, defective construction or subsequent failure of the project, any of which may result in serious injury to personnel and even destruction of the project.

The implementation of QA and QM procedures in no way diminishes the responsibility of suppliers and contractors to achieve quality. The emphasis of the approach is on eliminating errors and deficiencies by better quality planning, training and organisation, rather than by inspecting to find faulty products. The clarification of the responsibilities of the various parties involved is a key benefit of this approach.

Quality Management

The present way of doing things has tended to be the outcome of countless ad-hoc 'practical' answers to the incidental problems posed by the changes and urgent demands of our business. Such an approach has its functional costs in that too high a proportion of skilled and scarce resources are deployed in steering the production process itself; adjusting, re-blending, fixing problems, delays and holdups — day-to-day crisis management. Technical and managerial talent should be concentrated in preparation, planning and improvement — not sweeping up.

Quality management is a systematic way of guaranteeing that organised activities happen as they were planned. It is a management discipline concerned with preventing problems from occurring which, by changing attitudes and providing controls, make error prevention possible.

The foundation and spearhead for *improvement* by prevention is the quality management system driven by management at all levels.

Quality Assurance

The definition of quality assurance is as follows:

> 'All the planned and systematic actions that are necessary to provide confidence that a product or service will satisfy given requirements for quality'.

The current British Standard for Quality Assurance, (BS5750) is applicable to all organisations (industrial, commercial or governmental) which produce a product or provide a service intended to satisfy a user's needs or requirements. It is therefore applicable to all projects.

Quality Assurance for any project can be accomplished with the implementation of a quality system, which may be based either on a client's or contractor's existing quality system, or one specifically for a particular project.

The Standard (BS5750) provides a complete checklist of the areas which need to be included in a quality system. A description of the quality system is usually set out in the quality manual which references supporting work procedures and detailed job instruction. The quality system should be supported by a specific Quality Plan.

Obviously the scale of quality plan will depend on how geared the existing quality system is to project work. A contractor specialising in engineering, procurement and construction will have quality documentation more geared to the requirements of a particular project. In a complex client organisation, the Quality System will be more broad-based to encompass the multiple activities and the interfaces between them.

The Quality Plan will therefore highlight the specific requirements, the various sections of the Standard which need to be reinforced for the requirements of an engineering, procurement and construction type project.

Quality Control

The definition of Quality Control is as follows:

> 'The operational techniques and activities that are used to fulfil for quality'.

In order to ensure that the essential areas of Quality Control (QC) are adequately covered in a major construction project, it is necessary to split QC into the following activities:

Inspection Procedures

The final procedures required for the project should be identified and agreed well in advance of vendor selection, so that the equipment and material purchase orders can then clearly specify these requirements. If, for any reason, these requirements are agreed after equipment and material orders are placed, very significant additional costs and serious delays can occur.

Examples of inspection procedures are:

Non-destructive testing heat treatment, hydrotesting, fabrication and construction for all disciplines eg mechanical, civil, instruments and electrical.

Vendor Selection

Before vendors are short listed for purchase order bids, it is essential that the QA and QC experience of each potential vendor is assessed to ensure that the required level of quality can be achieved. Certification of the vendor's quality system by an accredited certification body may provide a useful indication of capability. Once bids are received from selected vendors, then a QC assessment of each bid should be carried out along with the technical assessment before the final selection is made.

Project Design

Quality Control of a project is maintained by carrying out detailed audits on drawings and equipment requisitions to ensure that full compliance with project specifications has been achieved.

Pre-Inspection Meeting

The pre-inspection meeting should be held at the purchase order placement stage in order to agree the inspection plan and any inspection hold points required for staged Quality Control. The detailed non-destructive testing techniques should also be agreed to ensure that no misunderstanding exists between purchaser and supplier.

Vendor Inspection

Quality Control of vendors' work throughout the duration of manufacture of equipment and materials is essential to ensure the desired level of quality.

Purchase orders should specify the level and extent of Quality Control to be carried out at the supplier's works. This QC activity may be carried out either by the supplier's own inspection department or by a nominated third party. However, it is advisable for the client to carry out its own QC audits on suppliers to ensure that the specified level of QC is being maintained.

These audits should vary from spot visits to 100 per cent coverage depending on type or order and problems encountered during manufacture.

Examples of QC audits cover N.D.T. results, visual examinations of fabricated plant, test certification, etc.

QC assessment of vendor design drawings is also carried out to ensure compliance with specification.

Goods Inwards Inspection

When equipment and materials arrive at the project stores it is essential to check that these items comply with specification and that no damage has occurred.

This activity can be achieved by setting up an inward goods inspection facility. This facility will check received goods before they are issued to site, thus ensuring full compliance with specification. The percentage of goods checked should be defined by the client, taking into account service and type of material. If any goods are found deficient, they should not be issued until either the deficiency is resolved or the goods are replaced.

Examples of deficiency found on inward goods inspection are:- incorrect certification, wrong material, damaged material, incorrect tolerances etc.

Site Inspection

Once the equipment and materials have been issued to site for construction, it is essential that Quality Control is maintained through to project completion to ensure compliance with project specifications.

The QC activity will cover all disciplines. Each part of each work activity should be checked to ensure that no significant re-work will be required owing to late discovery of defective construction work.

Project Reports

In order to improve the quality of future projects, it is an advantage to compile a project completion report which will record the project activities from design to completion. The report will highlight problems encountered and give recommendations for learning points which can then be included in future projects to improve quality.

CONCLUSION

The ultimate objective of project management must be to give client satisfaction. A major source of dissatisfaction is the failure, for whatever reason, to deliver a product of the quality required. Achieving the necessary quality is particularly difficult in the construction industry because of its fragmentation and division of functions; but if the proper systems are installed and operated by properly trained and motivated people, the difficulties can be overcome to the ultimate benefit of all concerned.

8 HEALTH AND SAFETY

VICTOR COLEMAN
Health & Safety Executive

All accidents are preventable. Evidence of this is found in the achievement of many millions of manhours worked without serious injury. Such achievement is possible only with a well thought out safety strategy and the determination to carry it out. Death and injury will not be prevented by simply expressing good intentions and hoping for the best.

In Britain, in a recent 10-year period, 1,500 people died and over half a million people had accidents which caused them to be off work for at least three days as a direct result of some kind of failure on construction sites. The toll of death and disability from occupational ill-health in the industry is less easy to estimate but is likely to be at least as bad.

Figures 8.1 and 8.2 give typical analysis of the principal immediate causes of construction accidents in a given year. This helps to focus attention on the kind of tragedies we are most concerned about. It will

Figure 8.1 *Construction Accidents 1986-87* (fatal and reported injuries to employees)*

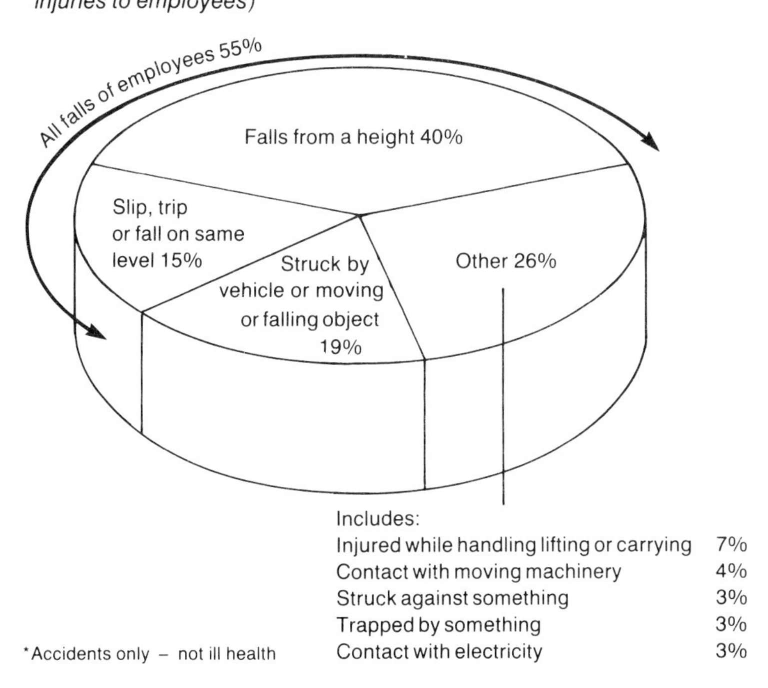

*Accidents only – not ill health

Figure 8.2 *Construction Accidents 1986-1987 (all reported injuries to employees)*

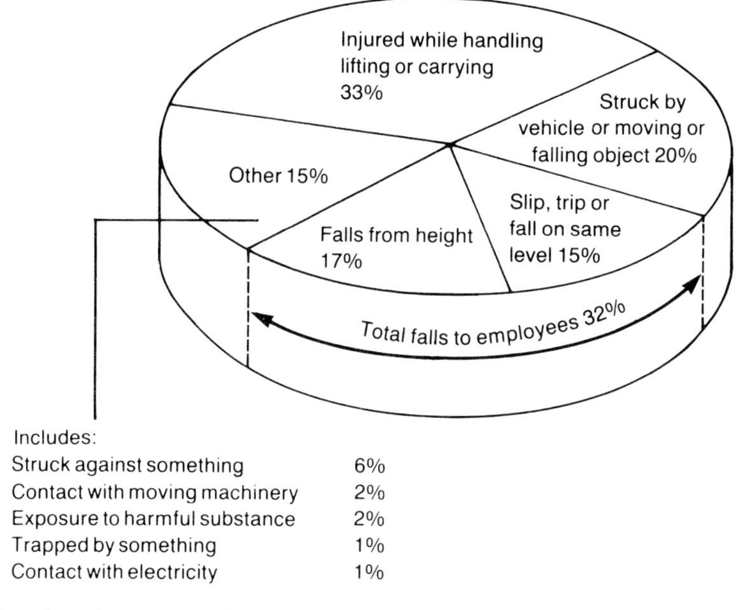

be clear that, above all, we need to tackle the problem of people falling which alone accounts for over half of all reported deaths and major injuries.

Each accident and each case of occupational ill-health has more than one cause. At a simplistic level we might say a scaffold collapsed because it was not tied to the building. But more importantly we ought to ask why was it not tied? Was the scaffold erection company competent? Who inspected the scaffold? Was the work of various contractors on site properly co-ordinated and controlled? Was the site properly managed?

The HSE estimates [Ref. 1] that nearly 90 per cent of all construction accidents leading to death could (and should) have been prevented, 70 per cent by positive *management* action. Nor is the problem of poor management control of health and safety hazards confined to smaller contractors and work on smaller sites. Nearly 30 per cent of all workers who died as a result of construction accidents in Britain in the 12-month period 1987/1988 were employees of (or they died on sites run by) one of the top 100 construction firms in the country. Clearly much more needs to be done to ensure safety. At a time when construction accounts for nearly a third of all people killed by work accidents in Britain, this has to be a high priority.

All contractors employing five or more people must have a written safety policy [Ref. 2]. Ensuring the proper management of health and safety on construction projects should never be seen as a burden or something rigidly to be separated from other aspects of management, or worse, something to be ignored! The principles outlined in this chapter will be already familiar to 'good' managers. A well-planned and well-run project will be both safe *and* efficient. It will save lives, injury, ill-health *and* money.

CASE STUDY

A series of accidents arising out of poor site conditions due to bad 'housekeeping' led an inspector to insist upon the contractor employing a labourer whose sole duty was to keep the site tidy, gangways clear etc. At a return visit the inspector was surprised to find the previously reluctant site manager jubilant. Not only had there been a reduction in accidents, but the saving on materials and the greater efficiency of moving materials around the site had paid for the labourer's wage *and* left a surplus.

WHAT MAKES FOR GOOD MANAGEMENT?

The efficient management of construction work builds on good *planning* and *commitment* by all concerned to do things right throughout the course of any contract. Whilst this has long been recognised by most parties to major construction projects, success in preventing accidents and occupational ill-health has been and continues to be variable. The Construction Industry Advisory Committee (CONIAC) which advises Britain's Health and Safety Commission (HSC) has recognised the need to produce straightforward and practical guides to 'Managing Health and Safety in Construction': the first dealing with principles and their application to main contractor/subcontractor sites [Ref. 3] was published in 1987; the second dealing with management contracting [Ref. 4] was published in 1988. These are essential guides to anyone involved in designing, planning or controlling work on construction projects of all sizes, but especially major construction projects.

CONIAC'S guidance makes a number of key points:

- ▲ The responsibilities for health and safety on each project should be clearly defined and reflected in contractual arrangements.
- ▲ The management of health and safety should be an integral part of the management of the work, and whoever is responsible for co-ordinating the activities of others on site should ensure that health and safety are effectively managed.
- ▲ Hazards should be anticipated, suitable plant and equipment

identified and someone made responsible for its provision and maintenance. Appropriate working methods need to be planned accordingly. Proper method statements are invaluable, providing proactive commitment and understanding. For example, when erecting steel frames, the use of preassembled structures can significantly reduce the need to work at height, the need to establish good hard-standing for mobile platforms can be specified along with the way fall arrest equipment is to be used.

- ▲ The design team should identify major factors which could affect health and safety and inform prospective contractors of them.
- ▲ Prospective contractors should not be selected or placed on tender lists unless they can show competence in the management of health and safety.
- ▲ Common, priceable items which are necessary for health and safety should be considered for inclusion in the contract documents.
- ▲ The organisation of site safety should be planned in detail, rules established and performance monitored routinely and by special safety audits where appropriate.
- ▲ Everyone involved in a construction project has certain responsibilities under health and safety legislation: clients and the professional advisers, management contractors, works contractors, subcontractors, and individual managers and workers.
- ▲ The most common type of site management framework is one based on the traditional arrangements involving a main contractor and a number of subcontractors. Figures 8.3 and 8.4 set out some key aspects of successful management for both parties.

The CONIAC guidance [Ref. 3] offers practical help in presenting a series of forums to provide a focus for discussion between main contractors and subcontractors.

Management contracting is becoming more common, especially for major construction projects. The management contractor is usually a construction company but does not normally carry out any direct construction work on site. The key difference between the many different types of management contracting is the extent of the contractor's contractual responsibility for design *and* construction work. Arrangements whereby the management contractor has access, at an early stage of the project, to design team information provide the opportunity for better integration and control of design and construction processes and so lay the groundwork for successful health and safety performance on site.

CONIAC guidance makes it clear that the management contractor has such a central role in the project that if he does not fulfil his health and safety responsibilities then no one else will! His main responsibilities will be:

- ▲ To help the design team identify information on major health and safety matters which would be passed on to the works contractors.
- ▲ To contribute construction health and safety expertise to the design team. This should lead to design features which are easier and safer to construct.
- ▲ To identify essential, separately priceable health and safety items (eg access scaffolding, edge protection and welfare facilities) which could be included in contracts with works contractors.
- ▲ To shortlist works contractors who may be invited to tender, taking account of their skills in managing health and safety.
- ▲ To draw up site safety rules and conditions, incorporating those of the client where appropriate, for inclusion in the contract documents with the works contractors.
- ▲ To consider site safety procedures, guidance notes and codes of practice which could be referred to in contract documents with works contractors.

***Figure 8.3** Main Contractor/Subcontractor Sites. Key Aspects of Management*

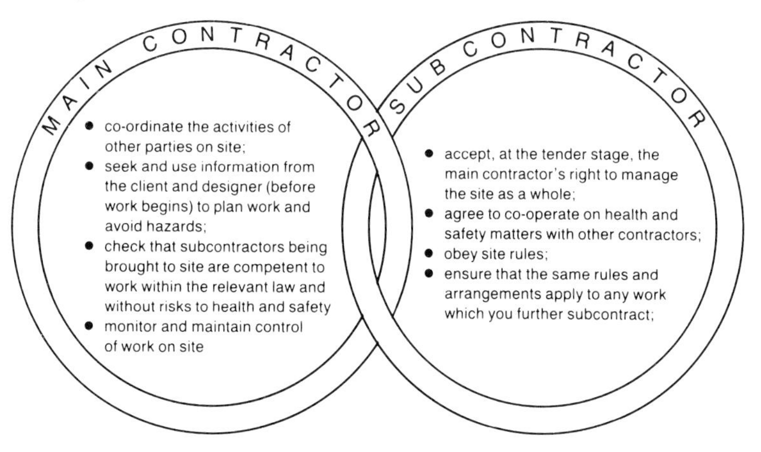

Figure 8.4 *Checklist for main contractor-subcontractor discussions*

Forms A

This checklist notes matters which should be discussed and agreed at an initial safety meeting between the main contractor and subcontractor. The tick boxes should be ticked when an item has been discussed and actions agreed. In addition, a separate note of the meeting should be made to record the agreed actions, comments and details required by these forms.

Contract Title

Name and address of subcontractor

Date and those present at the meeting

1 Information to give subcontractor(s)

Main contractor's safety policy ☐

Rules and conditions for subcontractors (Client's rules/ procedures (eg permits to work) ☐

List of external contracts (eg HMFI, local authority, emergency services etc) ☐

Names of main contractor's safety adviser(s) ☐

2 Information to obtain from subcontractor

Subcontractor's safety policy ☐

Names of subcontractor's safety advisers/supervisors ☐

Method statements ☐

3 Project meeting (attendance, frequency, content etc) ☐

4 Codes and standards

Agree those relevant to the work (some may have been based on the contract documents). ☐

5 Safety representation (nominations, arrangements etc) ☐

6 Training ☐

All employees should be given any necessary induction training/instructions on their arrival on site. Any employees required to do special tasks, eg forklift and dumper driving, banksmen and first aid duties should have appropriate training.

7 Sub subcontracting ☐

Arrangements to ensure safe methods and system of work when subcontractor intends further subcontracting

8 Accidents/diseases/ dangerous occurrences ☐

A death, specified major injury or condition should be reported immediately to the enforcing authority and the site manager. All injuries should be immediately recorded in the site accident book. Subcontractors should also record particulars for their own purposes. Forms F2508 or F2508 A, should be sent to the enforcing authority and copied to the site manager.

9 Hazardous conditions outside the control of the subcontractor ☐

The subcontractor should report to the site manager (or his site safety supervisor) hazardous conditions outside the control of the subcontractor, but which might expose his employees to risk, eg unsafe scaffolding, holes in floors or roofs, dark areas etc. The subcontractor should not allow employees to work in, or have access to such places until the hazard has been eliminated.

10 Welfare facilities/first aid/ occupational health and hygiene service ☐

State who provides them and note any exceptions. Arrangements for shared facilities to be confirmed on form F2202 by the contractor providing the services.

11 Housekeeping ☐

Arrangements for storing materials and clearing away waste materials.

12 Permits to work ☐

(a) Does the client's or other contractor's permit to work procedures apply to the subcontractor?
(b) if so, is the subcontractor aware of the requirements, and has he worked under this procedure before?
(c) details of any subcontractor's own permit to work procedures (eg electrical switchgear, substations etc) should be available. Are they?
(d) how to apply for a permit?
(e) who is issuing authority?
(f) what types of permit apply?
(g) how long does it apply?
(h) who can accept and sign off a permit?

13 Fire precautions ☐

(a) is spacing and construction of temporary site accommodations considered?
(b) is appropriate fire fighting equipment to be provided for all temporary accommodation, wherever liquified petroleum gases, highly flammable liquids or materials are stored or used and whenever welding/cutting is undertaken?
(c) have arrangements for raising the alarm and evacuation (where necessary) been made?

14 Emergency procedures ☐

Arrangements for raising the alarm, rescue and evacuation in event of an emergency other than fire, eg release of toxic fumes.

15 Access to site and site security ☐

(a) have site security arrangements been made?

(b) for road closures has there been liaison with authorities?
(c) is there safe access for vehicles and pedestrians?
(d) have arrangements for deliveries and unloading been made?

16 Measures to protect the public ☐

(a) has site fencing been considered?
(b) are scaffolding fans needed?
(c) has access to scaffolds etc been prevented?
(d) are holes and edges fenced/covered?

The management contractor is also responsible for setting up the site organisation for the management of health and safety taking into account the overall programme for the project, planned procedures, arrangements for co-ordination, liaison and communication, safety representatives' functions and arrangements for joint consultation, arrangements for monitoring site health and safety, arrangements for training, instruction and information, policy on the use of common facilities, plant and equipment, arrangements for record keeping and statutory examinations, external liaison, the responsibilities of sub-contractors and the responsibilities of individuals.

The management contractor should also ensure that works contractors are briefed about anticipated construction methods, site/design factors, relevant hazard, precautions, general site safety rules and conditions, and are clear about divisions of responsibility. Similarly the works contractors should inform the management contractor about possible hazards arising from their own activities.

He must satisfy himself that works contractors have made plans to carry out the work safely, have priced their bids accordingly and have the necessary resources. In particular he should ensure that works contractors produce method statements for high risk activities such as demolition, structural erection and work with asbestos, and he should assess those methods.

He should manage health and safety on site by co-ordinating activities, ensuring that planned procedures are implemented and monitor performance so that revised arrangements can be made as necessary. The management contractor should ensure that he does not become remote from day-to-day problems on site.

He should consider the creation of a joint safety committee operating on a site-wide basis and involving representatives of management and operatives from all contractors as well as make site-wide arrangements for emergencies, safe access, lighting etc where such matters are under his control.

THE IMPORTANCE OF COMMUNICATION

In any major project there need to be clear lines of communication between all the various parties on site, and information and problems will often need to be passed 'up' the structure — communication should not be a one-way flow. It is essential for subcontractors to nominate a person to deal on a day-to-day basis with the site manager. Worker representatives need to be identified and encouraged. It is especially important for the workforce to know where to go for a decision and action if a situation of imminent risk develops.

The detail of joint consultative arrangements on health and safety will vary according to whether there are to be organised trades union representatives on the site, and will be dependent on the size and complexity of the project. Where safety representatives are appointed by recognised trade unions, their employers should provide them with the facilities and assistance necessary to carry out their functions which include inspection, investigation, representation, and consultations on behalf of the workers they represent. The safety representatives may formally require a safety committee, which normally would operate on a site-wide basis. Such committees should play a major role in scrutinising health and safety arrangements and reviewing overall performance. The legal basis for this is set out in The Safety Representatives and Safety Committees Regulations 1977 [Ref. 5].

Even where there is no organised trades union representation on site it is still important to consult the workforce adequately and a safety committee may still be a useful way in which various parties can effectively carry out a number of their functions.

WHAT MAKES A GOOD MANAGER

Having looked at various aspects of management and organisation in terms of tasks, duties and functions, it is worth considering what qualities a contract manager, project manager, site manager etc ought to possess. In order to be regarded as competent with regard to safety a manager must have:

- ▲ adequate knowledge of the law and ways of avoiding accidents and occupational ill-health;
- ▲ the necessary authority,time and resources to achieve and maintain 'acceptable' conditions and systems of work on site;
- ▲ the will and ability to apply that knowledge.

This implies that 'managers' (at all levels) must get sufficient information and instruction to achieve knowledge of both what to do and how to go about it. They must be properly equipped by contractors and clients (by terms of contract or otherwise) to have authority (ie powers) and must have the personal qualities which are necessary (eg personal commitment, leadership and organisational abilities). The ability to motivate and control others is crucial and this can involve the promotion of incentive schemes and general 'propaganda' in addition to normal line management functions.

Depending on the type of job and complexity of the work, site managers etc will often not be able to acquire an encyclopaedic knowledge of health and safety law, and details of ways of avoiding all types of potential accidents and ill-health. In such cases they must have ready access to sources of specialist information [Ref. 6 and 7] and advice (eg, from the specialist safety adviser). Crucially their own knowledge must be such that they know when further information and/or advice is needed. No manager is likely to achieve all this without training.

THE ROLE OF THE SUPERVISOR

The supervisors, and particularly those in first line supervision, are key players in all aspects of construction including health and safety matters. Training of these people and getting their commitment is a vital and necessary step to achieving results on site. For any project to be successful it will be necessary to devote time and resources to this group.

TRAINING

Proper training is equally essential for good performance of managers and supervisors as well as in the construction trades. It is important to learn the right way to do any job. The right way will be the efficient way and the safe way. Simply picking up a job alongside people who have fallen into bad habits will serve to perpetuate problems of poor quality work and danger on site.

Everyone's training should cover basic requirements for health and safety on site. People in control of work, and especially those in charge of a site, must have a good working knowledge of what the law requires and of safe working practices. Safety advisors and safety supervisors need still greater knowledge.

THE SCOPE FOR IMPROVEMENT

The vast majority of accidents do not just happen. No matter what the specific 'cause' of any single accident or case of occupational ill-health a failure in 'management' (ie in planning, supervising, organisation or control) often lies at its root. Thus, safety is no accident. All construction projects need to be managed well. All parties have a role to play.

Effective action to tackle management and organisational problems ought to lead to improvement across the whole spectrum of construction health and safety problems. Time spent on improving the quality of project and site management is time well spent.

REFERENCES

1. *Blackspot Construction* HMSO, ISBN 0 11 883992 6.
2. *Guidance on the Implementation of Safety Policies* HSE, 1987 IAC L1.
3. *Managing Health and Safety in Construction: Principles and Application to Main Contractor/Subcontractor Projects* HSE CONIAC, 1987, ISBN 0 11 883918 7.
4. *Managing Health and Safety in Construction: Management Contracting* HSC CONIAC 1988, ISBN 0 11 883989 6.
5. *Safety Representatives and Safety Committees*, HMSO 1988, ISBN 0 11 883959 4.
6. HSE Catalogue *Construction: List of HSC/E References* from HSE enquiry points.
7. *Construction Safety* — 2 Volume loose leaf manual with updating service. Building Advisory Service Publication.

9 DESIGN AND ENGINEERING

JOHN COOKE
Davy McKee

Building and construction projects are generally unique. They require skilful management and a fundamental understanding of the design process. The design needs to be right first time, unlike that for an automobile or an aircraft. With these there is the opportunity to produce a prototype for testing and amendment prior to production.

SCOPE

It is imperative that both client and contractor have a common understanding of the contract definition. In simple terms, this includes a list of what is known, plus what is not yet defined, be it a design package or a fully engineered project.

Leading on from the definition, there is a need to ensure that all involved in the project understand its objectives and the means of achieving them. Contract (Job) Instructions are essential to all projects. This is a comprehensive document issued to all departments stating in precise terms the contract requirements including the programme, standards, co-ordination procedure, requirement and methods, eg CAD usage, modular construction, etc.

This instruction is the mechanism for ensuring all parties involved are aware of the exact needs of the project. Interpretation is therefore not left to any individual. The engineers who may have previously worked on very dissimilar projects need unambiguous statements of what is required.

COMPUTERISED OR MANUAL DESIGN

Prior to the design and engineering launch, clear judgements are needed on the use of CAD 2-D and/or 3-D.

The advantages and disadvantages of CAD are fineley balanced and its use has to be determined a on project specific basis

The advantages of 3-D high accuracy, through clash detection (both hard and soft) and other design checks within the software, plus excellent material take-offs leading to reduced field rework, are the critical factors for complex plants, modular construction and high accuracy.

The disadvantages of rigidity of information provided by CAD, the cost of CAD, limited skill availability and the possibility of incompatibility between the client and the contractor's systems, provide the rationale for selecting manual drafting and/or 2-D CAD as the project design strategy.

STAFFING

Neither client nor contracting companies can afford the luxury of under-utilised resources. Companies are forced to retain a core of skilled staff supported by agency personnel and subcontractors as work load demands. This can be an emotive topic as some clients believe that projects should be manned only by permanent staff. The reality is that many of the designer/draughtsmen can only be found on the agency market. They move as workload demands and the control and effective use of such staff is of enormous importance.

PROGRAMME

Programmes which are autocratically imposed may be a recipe for disaster. The overall programme together with the working schedules need to be built up of achievable elements. These elements need to be provided by the individuals concerned and the planner's task is to integrate these into a real programme. Commitment to the programme by the engineering disciplines is essential.

The planner needs to be totally aware of progress and delays affecting all disciplines. This is particularly true of the delays caused by receipt of data from licensor, vendor or another party. It is typical of industry that vendor data often tends to be both late and not the required quality. A pro-active approach is needed to maintain progress by adjusting sequences and priorities for engineering, procurement and subsequent construction. It is the joint responsibility of engineers and the planners to work closely together to maximise progress and minimise disruption to the schedule.

APPROVALS

Speed is of the essence in project execution and encourages engineers and designers to cut corners to maintain the all important schedule. 'Design-check-approve' is an essential approach for engineering and technical integrity cannot be put aside for the purpose of early completion.

Virtually all companies work to a matrix organisation but technical responsibility should always remain with the specialist line managers. Technical approval for documents and drawings should therefore be the responsibility of technical management, preferably external to the project team.

This is an area where QA audits can be particularly valuable in ensuring that the integrity of the project is maintained.

CONSTRUCTION-DRIVEN ENGINEERING

The objective of engineering is to provide a safe, operable project, on schedule, within budget and one that can be built. Occasionally, the fundamental objective is forgotten in the pressure of engineering to a tight schedule.

It is imperative that the resident construction manager designate is involved at the earliest stage of engineering. This provides the hands-on, committed experience that is so important.

Construction input to engineering is vital for a number of key reasons. Constructability needs to be designed into the plant from the earliest layout considerations. Construction personnel, therefore, need to be involved in plant layout particularly in the event of heavy lifts and/or if a modular concept is to be used. Construction personnel also need to be involved in work sequencing in order that deliverables from Engineering meet field needs. This is an area where construction, engineering, planning and materials control need to maintain a constant dialogue keeping the ultimate, overall objective in mind. Too often only lip service is paid to the term 'construction driven'.

PROGRESS

The use of manhours for quantifying progress is no real guide to achievement. Actual progress throughout the design and engineering phases needs to be rigorously monitored for assessing both real progress and adverse trends. Engineering productivity needs this close monitoring throughout, together with careful examination of trends.

A rigid approach to the issue of drawings and documents is probably the most effective method of progress measurement as this tends to provide a slightly pessimistic picture. Aspects affecting productivity such as poor vendor data flow can have significant impact on the schedule, if corrective action is not taken. An engineering task force or team working inefficiently is an expensive luxury.

Parallel to progress measurement, regular weekly progress meetings are needed. This is the forum through which lead engineers can understand each others' problems and perhaps become aware that they are causing difficulties for their colleagues. It is essential that progress meetings do not become the only occasion when engineering disciplines communicate with each other, and they must not become a forum for problem solving.

CO-ORDINATION

Involvement of the client, licensor, process and design engineering teams with major vendors and the field needs professional co-ordination. The job instructions need to provide a clear statement of each party's involvement and the requirements for co-ordination. This is particularly true in respect of reviews and approvals.

Uncontrolled client or licensor involvement in the engineering phase can be quite disastrous. The client clearly has the right to be involved but that involvement needs definition and understanding. The client who becomes involved as an integral part of the project engineering team can maximise efficiency during this phase. A team effort is more powerful than efforts of the individual contributors.

CHANGE MANAGEMENT

A subject which may be swept under the carpet is change management. Despite the good intentions of all parties, change may be required and is almost inevitable. However, rigorous attempts to assess real need are essential. The impact of change on the engineering programme is always detrimental, with the real effect in cost and time frequently and dramatically underestimated. Particularly invidious is a series of small changes, none of which is significant but where the collective impact is often serious.

The key questions for the project manager/engineer are: 'Does it work?' 'Is it safe?' It is vital only to permit change for need, not for want. Opinion engineering is a destructive force.

The only approach to the control of change is a highly structured formal system. Change alerts leading to review, impact assessment, resolution or redefinition, notification of either internal budget change or contract variation should be integral to the system. Discipline is necessary when change is occurring, however it is initiated. Attempting to engineer in an environment of continuous change is destructive; engineers and designers become demotivated and manhours are wasted. It is in clients' and contractors' interest to control and manage change and to minimise it.

COMPLETION

Completing the design and engineering phase is a difficult task and needs to be addressed almost as an individual project. Document finalisation is significantly affected and delayed by late delivery of certified vendor documents. At the appropriate time, about 80 per cent completion, each discipline should prepare a final check-list and effort should be directed towards effective completion on an item-by-item basis.

THE TEAM

An engineering team is a delicate flower that needs nurturing, and with care and effort it will blossom. Engineers are not only interested in producing specifications or drawings, they really do want to know how the overall project is progressing. Site photographs, a project newsheet, and informal gatherings all generate that important team atmosphere. The client and contractor working together to achieve synergy in completing engineering pays immense benefits to the overall project. This is not just desirable, it is essential. Without it the flower will wilt.

10 PROCUREMENT OF MATERIALS

TREVOR KENT
King Wilkinson Limited (a Babcock Company)

The primary objective of good procurement practice is 'to provide end-users with what they need, when they need it, at the lowest cost'. It begins with defining the requirements of the project, is followed by the selection of suppliers or subcontractors, and ends with the delivery of material at the destination.

To purchase material to specification from the best source, at the right price and delivery is a challenging task. To do this in a way that is completely auditable requires carefully thought out procedures, high standards of administration and a good deal of experience.

Suppliers, clients and financial institutions all require, in their different ways, security. Suppliers want to know that they stand a fair chance if they submit a tender and that their prices and documents will be treated in confidence. Clients and financiers want to see that their money is properly spent and that all decisions can be justified in accordance with laid down procedures.

It is the dual requirement of confidentiality and auditability which is at the heart of the procurement function and which in large measure determines the documentation procedures and working practices.

PROCUREMENT STRATEGY

The success or failure of a project is closely linked to the procurement strategy. If this is well thought out and competently run the project has an excellent chance of success.

Typically, a main contractor or owner will purchase much of the equipment and material directly from suppliers for free issue to subcontractors.

The first step in the preparation of the procurement strategy is the study of the prime contract. From this document the salient points to be included will emerge. The central concern is risk, both the quantum, where it falls, and how it can best be mitigated. It is the passing down of this risk to the many suppliers and subcontractors that constitutes one of the main contributions of the procurement function.

It is difficult to quantify all the risks involved in a project, but they can be grouped into categories and dealt with in one of a number of ways.

The first is financial; is the contracting party solvent? Reference of one of the credit agencies will normally elicit the information required. Bank references are another way, but these are seldom explicit and often require supporting information from other sources to get a complete understanding of the position. The financial structure of many companies is extremely complex so that even after diligent searching it is not clear how strong they are. If the contract is for the purchase of goods cash on delivery, there is little to worry about. If however, substantial progress payment is involved, there could be a problem in the event that the company goes into liquidation. It is a sensible precaution to ensure that the purchase order states that title to the goods passes to the buyer progressively as payments are made. This gives the right to remove partially finished items from a works and get them finished elsewhere. In the face of strikes, insolvency or natural disaster, this may be an unavoidable step.

Another tool in the buyer's hands is the Bond. This can be used for a number of purposes. The first is the Bid Bond where the buyer wishes to ensure that a party (normally a party who has prequalified and shown interest) will tender. The second is the Performance Bond which obliges the contractor to fulfil the terms of his contract on penalty of forfeiting the Bond. In some cases it may involve meeting a number of specific requirements, plant performances criteria, effluent discharge, product quality and many others.

Bid Bonds are normally for a specific sum, while Performance Bonds tend to be for a percentage of the contract value, commonly between five and 10 per cent, but can be for amounts exceeding the contract value. The bond is redeemable at the end of the contract which is often the expiry of the guarantee period. In some instances, buyers elect to retain a part of the contract price for a similar period instead of requesting a bond.

There are basically two types of bond: the On Demand Bond which is an irrevocable letter of credit that can be called at any time at the sole discretion of the buyer; or a Bond which can be called by the buyer only on producing corroborated evidence that he is entitled to the Bond.

A further precaution is the Parent Company Guarantee. This is employed when dealing with small companies which are members of large groups. The buyer will commonly ask the holding company to underwrite the contract.

A major risk in any project is delivery delay. This can disrupt the construction schedule and result in costs totally disproportionate to the cost of the late materials. This is often dealt with by asking suppliers to accept penalty clauses to ensure timely delivery. Penalty clauses are only enforceable if matched by equivalent bonus clauses. However, liquidated damages clauses which are often expressed as a percentage of the price forfeited for each day or week late can be enforced in their own right. These usually specify a maximum figure of five or 10 per cent of the contract price and in some cases a bonus payment is expressed in similar terms for early delivery.

Liquidated damages represent the best estimate of the parties as to the loss that would be suffered by the delay. The actual loss may be greater, but the existence of this clause effectively prevents the injured party from suing for damages at large. To this extent the clause also protects the vendor.

There is one further device that can be employed to facilitate timely delivery: stage payments against performance. Written into the contract schedule are 'milestones' corresponding to the achievement of certain pre-agreed stages of completion. If the stage is reached, the payment is made, if not it is delayed. It is important that the milestones are easily identifiable, such as delivery of castings, pressure testing of vessels, final performance, etc. If the stage is expressed as a proportion of the contract value there will be difficulties measuring the actual progress and endless debate as to what is equitable in the circumstances.

PROCUREMENT ORGANISATION

Procurement Departments can be organised in various ways and often report into their own organisation at different levels, depending on the way the company itself is organised and the status the function is given. In some companies there is a procurement director reporting to the Board; in others a procurement manager reporting to the director of operations. Typically, the procurement manager will be responsible for purchasing, expediting, inspection and shipping, and sometimes for subcontracts. In other organisations the inspection function may be managed by engineering and subcontracts managed by the construction department. There is no rigid structure.

Project procurement may be handled on a functional basis or within a project task force. Both systems have been successfully employed and it is a matter of company or client preference which system is adopted.

The functional system uses the in-house procurement department to perform the work and relies on the procurement manager to liaise with the project manager. The project based taskforce approach relies on a project procurement manager to oversee the procurement function. In some cases he may have one or two specialist buyers directly reporting to him, but normally he will use the department personnel as the resource for the job.

Whether or not purchasing is organised as a separate specialism or as an integral part of project management depends very much on the nature of the purchases to be made. If the goods and services to be procured are an integral part of one or of a limited number of major projects, it is natural that the purchasing should be organised in a way that is integral to the project management operation as a whole. In that situation one would in fact expect purchasing staff to report to those heading the project management team. If however the goods and services required are also required by a number of other projects, and are not highly individual in terms of specification, it is more natural that procurement should be done by a separate organisation reporting to a director of procurement. In that situation, the project management team would secure the services of a separate procurement organisation to purchase the necessary goods.

The benefits and drawbacks of these organisational structures probably depend more on the personality of the people involved than any other single factor. There is, however, one aspect that needs to be watched very closely by the procurement manager if a taskforce system is employed. While the project procurement manager is responsible to the project manager for implementing the work, he is functionally responsible to the procurement manager. There is a tendency when working closely with other members of a project team for verbal communication to supplant the written word which strikes at the root of good practice and particularly at maintaining an auditable system.

PROCUREMENT PROCEDURES AND REPORTS

One of the earliest tasks for procurement is the preparation of a procedure and control document which will take into account the client and contractor's own procedures together with any specific requirement for the project. Typically, procedures will address: supplier lists, commercial terms & conditions, requisitioning, enquiries, bid evaluations and authorisations, purchase orders, vendor documentary requirements, expediting, inspection/QA requirements, shipping and reporting and controls.

The evaluation and selection of potential suppliers and subcontractors for inclusion on the supplier's list for a project requires careful attention and normally includes companies from the contractor's standard vendor listing together with preferred client suppliers (if any). The location of the site, currency of purchase, quality constraints and any special engineering standards which may be required have all to be considered. When required, pre-qualification questionnaires are issued to establish the technical and commercial competence of potential suppliers, and as in all dealings with suppliers, assessments must be objective and conform to the highest ethical standards. The objective is to establish a comprehensive list which will ensure commercial competition from suppliers, each having a fair and equal chance of securing contracts from enquiries issued.

The commercial Terms & Conditions used can be a hybrid comprising the contractor's normal trading terms, industry accepted standard forms of contract and any additional terms and conditions deriving from the prime contract with the client. For major equipment orders and installation subcontracts, 'back-to-back' requirements on suppliers might be preferred.

A detailed treatment of terms and conditions used in heavy engineering contracts is included elsewhere in this book. However, from a procurement point of view, there is much to be said for using standard terms and conditions where possible. Models such as IMechE, ICE, and FIDIC, are widely employed in industry, and contractors know how to quote against them. Using novel forms gives rise to uncertainty which increases the contingency built into the bid price.

To ensure effective control, all requests for quotation and purchase must be initiated through a properly authorised requisitioning procedure controlled by the Project Manager. Engineering personnel are responsible for the preparation of detailed technical specifications which should include reference to all relevant specifications, standards and drawings necessary to describe completely the items to be supplied. They should not include extraneous commercial or contractual terms and conditions.

For packaged units, such as water treatment plants, desalters or lubrication consols, co-ordination of the technical scope may need input from several engineering disciplines. The object is to issue one combined specification for the whole package, with each engineering discipline contributing its own particular requirements.

Many packaged units contain a great deal of proprietary technology which may not conform to the standards of the project as a whole. In order to keep the price as competitive as possible, it may be decided that certain concessions are acceptable. This is an area where procurement and engineering need to work closely together to achieve the best results.

Bulk material requirements, piping, electrical, instrumentation and civil items are compiled with reference to engineering bills of material and parts lists on drawings. The volume and variety of bulk item requirements, their relative importance during the construction phases of the project, and the necessity from an overall 'control' point of view to manage this function from take-off to orderly storage at site, has encouraged the development of computerised systems. These systems may be integrated into computer-based requisitioning and purchasing procedures, thus forming a comprehensive material management system which in one form or another is now widely adopted throughout the contracting industry.

It is impossible in a short review to cover all procurement activities, but one which on occasion gives rise to particular difficulties should be mentioned. Single tender action occurs when there is only one supplier capable, for whatever reason, of fulfilling the contract. Clients sometimes nominate a supplier because they wish to standardise their plants, or feel that for safety or technical reasons they are better served by one rather than another. In some such cases, it can be a major exercise investigating the suppliers' infrastructure to obtain detailed cost information in order that a proper evaluation of the tender can be made. It would be exceptional to go to these lengths for normal commercial contracts, but it is by no means unknown in the nuclear and defence industries.

Enquiries for purchase initiated by the project authorised requisitions, comprise a full description, together with any necessary technical attachments, to which procurement adds the commercial terms and conditions and normally the requirements for interim and final vendor data and documentation. All enquiries should be issued simultaneously to suppliers from the agreed suppliers' list indicating bid return dates and any submission requirements.

If all goes well, no action is required until the tenders are received. However, it is not unknown for suppliers to request an extension of the bidding time. If a number of good companies on the list are capable of delivering their bids on time it may be unnecessary to grant the request. If there are only three, however, it may be prudent to do so if the delay can be accommodated within the overall project schedule. In cases such as these, all bidders must be treated equally and if an extension is granted to one, it must be given to all bidders.

Bids arriving before the closing date are retained with procurement until opened. The technical part is distributed to the appropriate engineering group to check conformity with the specification and the commercial part retained within procurement for analysis. All the evaluations are finally presented together on a bid assessment spreadsheet. Each technically acceptable supplier will be recorded as having conformed or not against a comprehensive list of questions concerning his terms of supply.

Often it is necessary to go back to suppliers to clarify their offers. While this is common practice, care needs to be taken that this is not used as an opportunity to rebid.

Review meetings are held if necessary to agree on an acceptable technical/commercial recommendation to be offered for approval to either project management or client representative. The bid evaluation will reflect any pre-order contractual negotiations held with prospective suppliers.

The 'purchasing' cycle of procurement normally ends with the purchase order award and the issue of confirming documentation covering the supply. Responsibility for the monitoring of progress, and quality is borne by the expediting and inspection groups.

EXPEDITING

If suppliers always performed to their promised timetable of events as agreed at order commitment, there would be no need for expediting. Unfortunately, in reality, the monitoring of progress through manufacture, fabrication, packaging and shipment of project purchases remains a critical element of procurement, particularly when overall project schedules demand timely material deliveries. The procurement documents covering expediting will indicate, depending on the criticality of the materials, the level and frequency of follow up and type of expediting required. Routine desk expediting may be carried out by telephone, facsimile or telex for non-critical items. For critical items, works visits are essential. This applies particularly to packaged plant and large items of mechanical equipment such as centrifugal compressors and gas turbines. In many cases, resident inspector/expeditors are assigned for the duration of the contract. Whatever method of expediting is used, the results must be recorded and reported in a material status report so that any actual or potential problems which might affect scheduled requirements can be readily identified and timely corrective action taken. Expediting frequency is normally increased as agreed delivery dates approach. This ensures that maximum supplier effort is exerted to meet project requirements. Expediting is not complete until receipt of all ordered items at the agreed delivery point has occurred.

INSPECTION

The importance of monitoring quality to ensure performance to design specifications requires the establishment of an overall inspection schedule at the outset of the project. The inspection schedule defines the level of inspection required using a grading system associated with type of material and criticality. Critical items require careful inspection and monitoring at all stages of manufacture against an agreed quality plan, whilst other materials (typically bulk material items) may only require a final check of test certificates. In addition to the detailed inspection of equipment and materials during manufacture, the inspector should witness all tests called for, check test certification, weld procedure radiography (and welder's qualifications if necessary) as well as final painting, marking for transportation, packaging and loading. An Inspection Release Note is normally issued and signed by the inspector when he is satisfied that all requirements have been successfully complied with and the material cleared for transportation.

Careful co-ordination of inspection and expediting activities are necessary to avoid duplication of effort. Inspectors are perfectly capable of expediting material and while it would be wasteful to use technically trained personnel permanently in this function, it is quite common for them to do a certain amount of expediting during their works visits.

SHIPPING

For major projects, the logistics involved in the co-ordination and control of the movement of materials and equipment from point of manufacture to the project site are often complex, particularly when the construction site is in a remote location. Careful investigation of possible transport methods, routes to the project site, and data concerning the capacity of ports of export and import including craneage is required. Equally important is a thorough knowledge of the documentation requirements, shipping, instructions, export and import regulations, tariffs and release notes which may be involved.

Often a detailed route survey is required to fully explore the logistics required particularly where the movement of large 'out of gauge' items are involved.

Consolidation of project materials either before shipping to overseas sites in source country marshalling yards, or at collection yards within the country where the project is to be constructed may also be necessary. It is usual, unless there is a heavy continuous workload within the transport section of the contractor's organisation, to utilise the services of external specialist shipping and forwarding agents to arrange for the movement of goods. In-house expertise is used to co-ordinate, negotiate and administer the transport requirements for the project.

REFERENCE

Procurement in the Process Industry — Albert Lester and Anthony Benning, Butterworth 1989.

Initial stage of Canary Wharf project

Construction of Brent C oil platform

(Below, left)
Lifting a pre-fabricated module into place

(Below, right)
Pre-assembled regenerator vessel prior to erection at Shell's Stanlow Site (planning for this vessel movement began 21 months earlier)

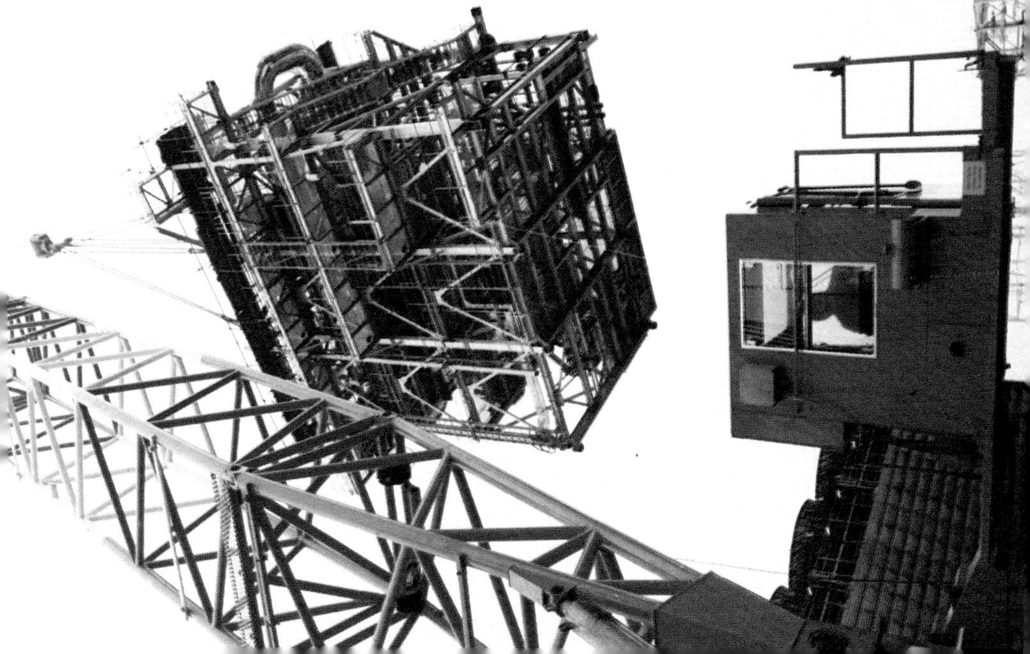

Lifting a pre-assembled regenerator vessel into place

(Below, left)
Rotary skip washing machine — British Nuclear Fuels' POND 5 project at Sellafield, Cumbria

(Below, right)
Installation of a 660mw turbo generator at Drax B Power Station

Long Residue Fluid Catalytic Cracker at Shell Stanlow (before completion)

Long Residue Fluid Catalytic Cracker at Shell Stanlow (after completion)

Phase 3 of the Broadgate Project, London

11 MANAGEMENT OF SUBCONTRACTORS

RICHARD WEEKS
M W Kellogg Limited

Work may be subcontracted for a number of reasons. Indeed the form of contract between client and main contractor might dictate that *all* work is subcontracted.

Other than where the main contractor is appointed as a managing contractor work might be subcontracted for any of the following reasons:

- ▲ To obtain access to specialist expertise, skills, knowledge, equipment, personnel.
- ▲ Where it might be politically undesirable to set up a construction organisation.
- ▲ Where, in an overseas location there are well established, capable, low overhead indigenous contractors.
- ▲ Where it is desirable and beneficial to let a complete design, supply and install subcontract, or concentrate the responsibility, or avoid abrogating design or supply warranties.
- ▲ Where the main contractor lacks expertise in a particular aspect or discipline of design or direct labour construction.
- ▲ To obtain the best resources available in the marketplace.

At one time it was commonplace for a main contractor to tackle all aspects of the works on a large construction project. It is however now the common practice to subcontract to contractors who specialise in certain types of work or particular disciplines (eg civil engineering, mechanical, electrical etc), or the provision of specialised services (eg brickwork, radiography, scaffolding etc). In subcontracting the work, the main contractor seeks to harness and co-ordinate the best available expertise in terms of management and supervision, labour, and the most suitable (and often specialised) construction plant and equipment.

THE SUBCONTRACT PLAN

The decision on what work is to be subcontracted, and the division of the main contractor's works into subcontract packages is an important decision, and is fundamental to the successful organisation and management of the project. It should therefore be taken at the earliest possible stage of the project. It is important that what is intended to be subcontracted is planned and everyone knows precisely the limits of responsibility for each subcontract.

Factors to be taken into account in the subcontract plan include whether individual subcontractors will be used for:

- 'Turnkey' construction through to mechanical completion.
- The provision of labour only by the subcontractor with some or all materials and equipment being provided by the main contractor.
- Discrete areas of work to a specialist subcontractor such as the fabrication of pipe (for subsequent free issue to an erection subcontractor), the supply and erection of furnaces, the complete building services system etc.
- Conventionally built project in-situ or in the form of skid-mounted units or pre-assembled units (PAUs).
- What element of design and supply if appropriate will be included in the subcontract?
- Whether the works will comprise only the erection portion or will include commissioning?
- Whether they will include the need for the supervision of erection by the suppliers of equipment such as compressors or other package plant?

SELECTION AND APPRAISAL OF SUBCONTRACTORS

A list of competent subcontractors capable of performing the subcontracts identified in the contract plan should be compiled at an early stage in the project. These subcontractors should then be invited to prequalify for inclusion on the tender schedule.

Failure to understand adequately the capabilities and limitations of potential subcontractors may lead to serious problems later.

Following are some of the items which should be addressed in a prequalification questionnaire:

- Previous experience and ability to perform works of the type and size required.
- The adequacy of labour, supervision and construction plant resources that will be available during the contract period.
- The past safety record and adequacy of safety policies and procedures.

- ▲ The ability to plan the works and maintain progress and productivity.
- ▲ The policy and procedures for quality assurance and control.
- ▲ What work would be sub-let, if any, and to whom.
- ▲ The policy and procedures with respect to industrial relations.
- ▲ The company's financial backing and capability.
- ▲ The experience and adequacy of the management and supervision proposed for the project. In particular it is important that the Site Manager satisfies the requirements because it is he who can make or break the job.

MEASURING AND REPORTING

On large multi-discipline projects, the measurement of progress achieved is a very subjective exercise, even with the most sophisticated progress measurement systems. It is an area which offers the potential for argument and dispute between the subcontractor and main contractor. If the subcontractor's payment depends upon progress achieved, he will then naturally report whatever progress he can get away with. If the measurement is not adequately verified by the main contractor, it will not only lead to the subcontractor being paid for progress he has not achieved, but will also paint a false picture of the project's progress. It will also lead to interface problems with subsequent subcontractors and late reaction by management (both subcontractor and main contractor) to bring the works back on schedule or to minimise the effects of delays.

As with progress measurement, the level of productivity achieved by a subcontractor's labour force is often a subject for debate. Normally the subject is not raised by the subcontractor until the later stages of the job when detailed calculations and graphs are often produced to show how he was prevented from achieving his planned productivity. It is therefore a subject which should be addressed by the main contractor at the very outset of the works. It is important that accurate records of the actual manhours expended on the works are maintained by both the subcontractor and main contractor. Regular (weekly, bi-weekly or monthly) comparisons of planned and actual productivity can then be properly determined and trends measured.

Most subcontractors will resist providing the main contractor with manhour expenditure records as the information contained can be used to the subcontractor's disadvantage when claims are submitted. It is essential therefore, that a specific contractual requirement is made for their provision. Relying on standard clauses such as 'providing such information as the main contractor shall require' will not generally be adequate for the purpose.

Equally as important as the reporting of progress and productivity is the reporting of industrial relations problems. It is vital to the project for problem areas or potential disputes to be identified as early as possible and for the appropriate attention to be paid to ensure their early resolution. Contractual requirement for subcontractors to report industrial relations problems to the main contractor immediately should be included.

AUTHORISING CHANGE

There are few subcontracts in the industry which go through their life without change. The reasons for the change can be many and varied but whatever the reason the change requires agreement and authorisation in some form. The authorisation process usually starts with the issue by the main contractor of a request to the subcontractor to perform some additional or varied work. The request can be a simple letter, but more usually is in a standardised form known as a site instruction, work order, field instruction or similar title.

The subcontractor is usually required to price the request and notify what effect the change would have on the schedule. The main contractor has the option of accepting the subcontractor's quotation, negotiating a more favourable price, or cancelling the request.

Alternatively, the works may be priced on a unit rate basis, dayworks or some similar means, whereby the quantity of work performed can be established later and paid for at pre-determined rates or similar methods of payment. Whatever the method of payment the subcontractor should receive written authorisation before proceeding with any such additional works. It is an act of folly to allow a subcontractor to proceed with a change to the contract, however small, without first agreeing in writing the extent of the change, the cost or means of calculating the cost and the effect on the contract schedule.

It is usual to accumulate on a regular (monthly) basis all the agreed instructions to the subcontractor and incorporate these into a formal change to the subcontract in the form of a change order.

The following illustrates the main categories of changes to a subcontract:

- ▲ Changes to drawings
- ▲ Changes to specifications
- ▲ Revision of schedule
- ▲ Revision of quantities or units
- ▲ Changes in contractual conditions or provisions
- ▲ Changes or substitutions in materials

▲ Methods of construction imposed on the subcontractor

▲ Backcharges

▲ Penalty/bonus variations of contract price

▲ Insurance/claim settlements

▲ Idle time.

HANDLING CLAIMS

Claims from subcontractors can be emotive issues from both the subcontractor's and the main contractor's point of view and often involve claims for significant sums of money, whether justifiable or not. Claims can be simply defined as demands for payment by the subcontractor for costs, losses or expenses incurred due to reasons outside the subcontractor's control. Whether the reasons for these additional costs are within or outside the subcontractor's control is usually the central area of debate.

While subcontractors' claims should not necessarily be invited, it is good policy to obtain confirmation from the subcontractor during the regular progress meetings that there are no claims outstanding. This is better than leaving matters to the end of the job when all the relevant people are dispersing, records have disappeared and memories are beginning to fade. The relevant documentation required in defence of claims should be collected and assessed currently with the job.

Claims should be reviewed on a purely contractual basis. Generally, the contract lays down the conditions under which claims may be submitted. Where there are grounds for sympathetic consideration, these may be separately identified in order that they may be considered when the level of the offer to the subcontractor is being established. It is usual to prepare a negotiating strategy in writing and to set the negotiating limits prior to any meeting with the subcontractor.

The penalty for failing to agree and settle a claim is usually expensive. Frequently, the dispute will proceed down the road to arbitration or go before the courts. Whichever route is taken, considerable expense and expenditure of management time is inevitable for both parties, irrespective of the outcome.

THE MANAGEMENT AND RESPONSIBILITY FOR SUBCONTRACTS

The ultimate responsibility clearly lies with the Project Manager. However, there are varying viewpoints as to which department within the main contractor's organisation should be responsible for subcontract management. Some main contractors have a separate Contracts Department which handles pre-contract operations and obtains input from the construction and engineering departments as required. On award of contract, the responsibility for the subcontract is passed from the contracts department to the construction department. Some companies place the responsibility for pre-contract operations with their procurement department. Responsibility eventually passes to construction on mobilisation of the subcontractor at site. Other organisations place the total responsibility within the construction department. As the end-user of the subcontracts and the department which has a major input into the subcontracts, the construction department is probably the best option. The construction department plays an important role in projects which are fast track and need to be construction driven. The required delivery dates should be set by construction as the ultimate user of the drawings, data and material provided by the other departments. Having the construction department responsible for subcontracting means the responsibility for each subcontract remains within one department from preparation of the enquiry document to final close out. The construction department is able to maintain continuity of personnel from pre-contract to field administration which obviates the need for 'handovers' from main office to site personnel. This is achieved by assigning the construction subcontracts manager/administrator into the head office taskforce in the very early phases of the project.

SITE INTERFACES

An area with great potential for dispute on a construction site is the interface of the works of different subcontractors. When establishing the subcontract plan it is therefore important to minimise the interfaces between subcontractors.

One method of reducing interfaces is to include responsibility for potential interface difficulties within the same subcontract. For example, if the underground earthing system is included in the civil subcontract a potential interface problem with the laying of paving is removed. However care will need to be taken to ensure precedents and practices align with the overall requirements.

It is important to make it quite clear during pre-contract discussions with the subcontractor what interfaces can be expected. Set down specific milestone dates to achieve, complete and vacate specific areas, and make the consequence of failure to meet these dates known.

It is important that the subcontract terms include a clear requirement for subcontractors to co-operate with others working on the project.

LARGE VERSUS SMALL

Should a project have lots of small subcontracts or fewer larger ones? As a general principle it is preferable to minimise the number of subcontractors on site. The fewer subcontractors you have means fewer enquiries to prepare and issue, fewer contracts to award and administer, fewer subcontractors (though not necessarily fewer men) working within a restricted area, and fewer interface problems.

However the overriding criterion can be the effect on site productivity of having just one big subcontract. Experience shows that maximum productivity will usually be obtained with a peak subcontractor labour force of 150 men. Interface problems can become irrelevant when compared with the loss of productivity and time overruns that can result from a subcontractor's management overstretched with more than this number.

The other aspect which needs considering is the project's vulnerability. Problems with small subcontractors are easier to handle than those of the larger contracts. Commitment of the men themselves to their employers will be better in the smaller unit.

REFERENCE

The Changing Role of Specialist and Trade Contractors by Colin Gray and Roger Flanagan, 1989 CIOB.

12 CONSTRUCTION

JOHN LING
Ling Management Consultants

Construction work is probably one of the most absorbing occupations. It offers a wide range of interesting engineering and management problems to be solved. This variety is introduced by the many types of project being built, the types of 'plant', their varied locations, the changing workforce, and the changing nature of the project as it progresses. This interest probably goes a long way to overcoming the many difficulties with which construction personnel have to cope on site.

EFFECTIVE MANAGEMENT AND SUPERVISION

Providing leadership, motivation and effective management, and supervision of the workforce is the major problem in the construction of projects. Therefore consider the following. Are you providing:

- ▲ training in leadership skills for all levels of management and supervision;
- ▲ motivation in the appropriate form of praise, reward and encouragement;
- ▲ communications whether directly face to face, one-to-one in group meetings or via a people-orientated company or on site newspaper;
- ▲ for the unsocial aspects of the task by ensuring that personnel get adequate breaks;
- ▲ for the effects of an ageing population and identifying and taking account of the value and advantages of experience;
- ▲ for training and retraining in management and engineering skills and techniques;
- ▲ adequate induction into company systems and procedures for new recruits and in the peculiarities of particular projects for all;
- ▲ a clear chain of command with a minimum of reporting levels and without multiple-head reporting;
- ▲ for responsibility and authority for management and technical decisions to be devolved to the lowest possible level;
- ▲ for the first line supervisor to have a clear understanding of his

responsibilities for motivating his team and of his authority to carry out the task?

THE EFFECTIVE WORKFORCE

Many of the items in the previous section apply to the men and women at the workface. However, experienced tradesmen have skills and knowledge of the business which is often overlooked. Solutions are often imposed on workers where a better, more economic, solution could well come from the men themselves. Analysis of the work content of a task using standard work norms does not necessarily provide the cheapest and quickest result. The man doing the task will often provide a better solution. Are you therefore:

- ▲ utilising untapped skills within your workforce;
- ▲ developing multi-skill craftsmen;
- ▲ deskilling many of the tasks that do not require the application of precision work;
- ▲ establishing a productive environment;
- ▲ providing motivation/involvement/communication/consultation;
- ▲ providing training and retraining in the latest techniques of construction;
- ▲ providing for the role of the unions in overcoming shop floor resistance to change;
- ▲ taking adequate steps against demotivation through an excessive level of design changes, poor supervision, inadequate instruction, poor facilities and poor working environment;
- ▲ taking adequate steps to neutralise the effects of militants;
- ▲ applying appropriate terms and conditions of employment for the particular working environment;
- ▲ taking every possible step to improve working conditions;
- ▲ providing adequate systems for the recording, measurement and control of absenteeism?

CONSTRAINTS ON EFFECTIVE PERFORMANCE

Criticism of the performance of construction projects is often directed towards site activities and inadequacies. Strikes, accidents, go-slows, high levels of absenteeism are high profile, highly reportable events.

However, they are often secondary causes of low productivity and delays. External factors have a far greater impact on construction performance than is generally accepted.

EXTERNAL CONSTRAINTS

Have you considered that inadequate, poor or late design cannot be cheaply rectified in the field. Dependent upon when they occur, correcting design shortcomings can cost twice to ten times as much as the original work through demotivation, disruption and delay.

Late materials and in particular out of sequence delivery of bulk materials (eg pipe spools without valves or supports, cables which cannot be terminated etc) are extremely disruptive to the effective organisation of work and can result in construction cost increases far in excess of the saving in value of materials bought from cheaper but unreliable suppliers.

Contesting onerous contract terms can divert management effort away from the essential tasks to an extent that far outweighs their presumed benefits to the client.

Ineffective and inappropriate client management organisations and personnel, especially those that require high levels of consultation and reporting, will be distracting and diversionary.

Inappropriate styles of contractor's management (eg using personnel with management contracting experience to run direct labour contracts, or those with reimbursable contracting experience to run hard price contracts) are a recipe for failure.

Unrealistic QA requirements can add to costs and cause delays without any enhancement of the project, while an unrealistic programme, (eg. one that sets impossible completion dates), will divert management efforts in countering its effects and possibly lead to uneconomic levels of labour being employed with consequent disruptive effects (see Figures 12.1 & 12.2).

An inadequate budget, perhaps resulting from poor scope definition, leading to ad hoc cost cutting or attempts to let underpriced subcontracts will quickly demotivate management and labour and have the opposite effect to that required (ie it will almost certainly push up costs and delay completion).

ON-SITE CONSTRAINTS

Some of the causes and effects of site generated delays and low production are:

Figure 12.1 *Realistic Construction Periods*

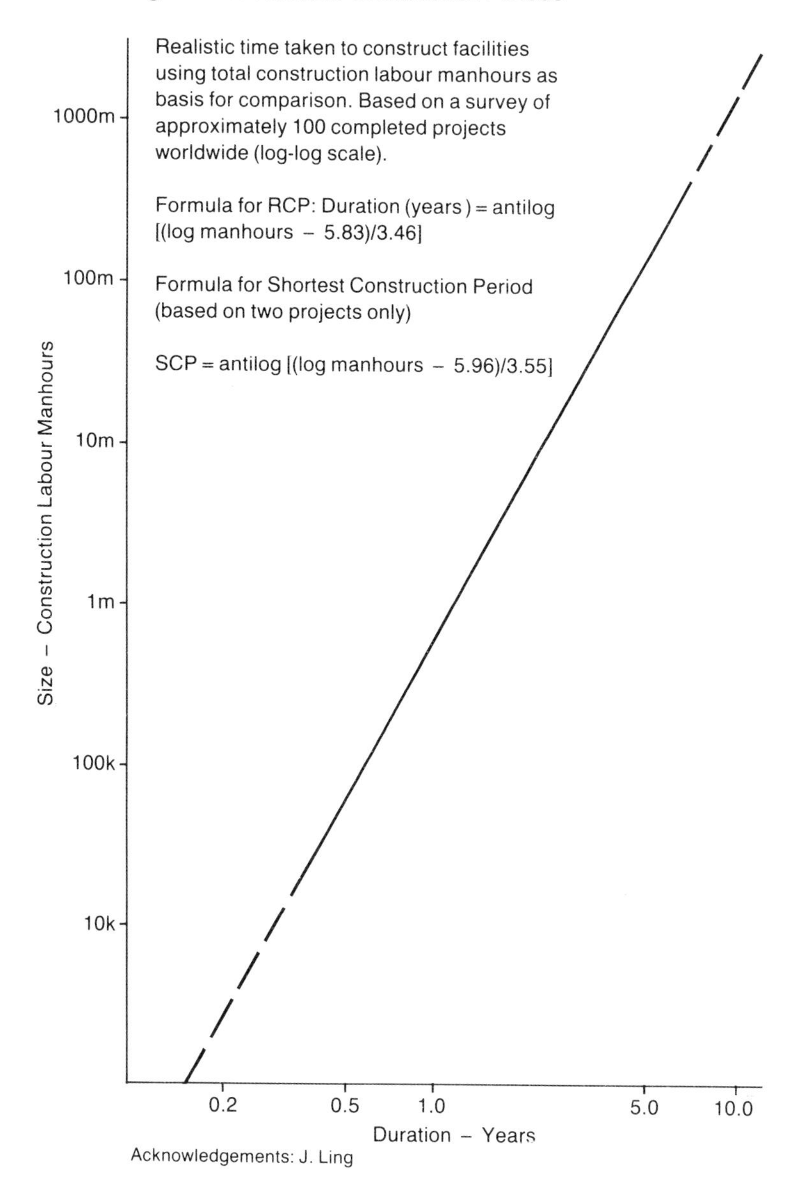

Realistic time taken to construct facilities using total construction labour manhours as basis for comparison. Based on a survey of approximately 100 completed projects worldwide (log-log scale).
Formula for RCP: Duration (years) = antilog [(log manhours – 5.83)/3.46]
Formula for Shortest Construction Period (based on two projects only)
SCP = antilog [(log manhours – 5.96)/3.55]
1000m
100m
10m
1m
100k
10k
Size – Construction Labour Manhours
0.2
0.5
1.0
5.0
10.0
Duration – Years

Acknowledgements: J. Ling

Figure 12.2 *Deterioration in Performance with Increasing Numbers Employed*

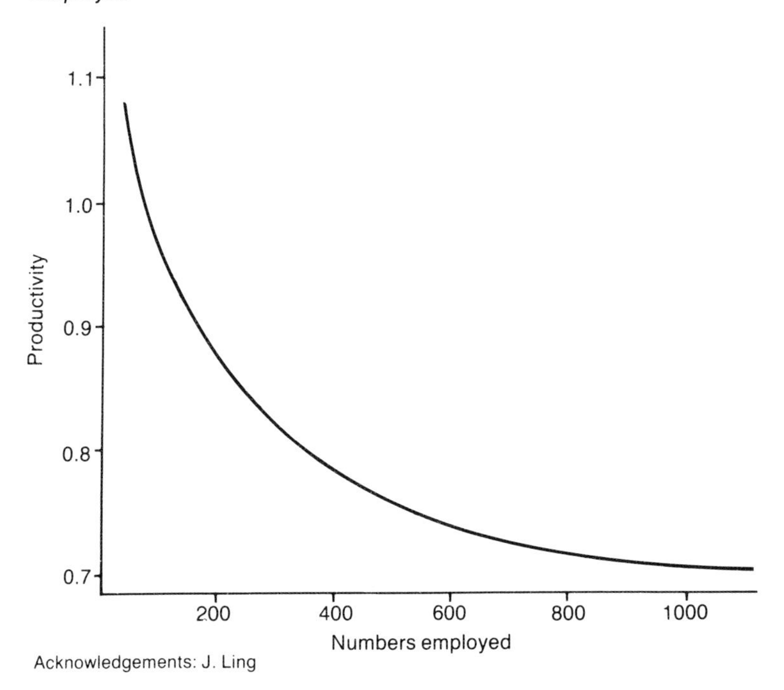

Acknowledgements: J. Ling

- **Inclement weather.** The cost of providing totally enclosing weather protection systems can rarely be recovered by reduced labour costs. Their justification lies in reducing the construction period to the ultimate benefit of the end user. Partial weather protection can be inexpensive but very effective.
- **Skill shortages** can lead to under resourcing and therefore delays with resultant low productivity
- **Working environment.** A poor environment will cause havoc with productivity. Early and adequate expenditure on hardstanding and roads is imperative, it reduces muddy conditions and improves safety. The cost of repairing damaged surfaces is minute compared with the value of savings made elsewhere.
- **Site layout.** Offices, mess huts, canteens, toilets, stores etc located too far from the workface add considerably to construction costs. Sub offices, mess huts, satellite stores, all located within minutes of the job, reduce unproductive walking time. Bussing large work forces to a main canteen can rarely be justified.

- ▲ **Supervision.** Inadequately trained, remunerated and motivated management & supervision cannot be expected to tackle and resolve productively the multiplicity of problems arising on the average site.
- ▲ **Demotivated workforce.** Lack of leadership and clear instructions and a poorly organised site will lead to a demotivated workforce and will exacerbate the demotivating effect of previously mentioned external influences.
- ▲ **Unco-operative labour.** Occasionally (though less frequently than the media would have us believe) militant and unco-operative labour may be a problem. Normally full time trades union officials will co-operate with management in helping to neutralise their effects, as long as effective management is otherwise being exercised.
- ▲ **Material control.** Properly organised material control systems oriented towards having materials ready when required and not provided at the last minute will save time. Queues outside stores buildings are a sign of poor materials control systems.
- ▲ **Document control.** A key requirement of any document control system is to ensure that superseded drawings and specifications are withdrawn quickly at all levels. Effective QA systems will normally take care of this and avoid rework.
- ▲ **Planning systems** pay for themselves and are an aid to improved productivity only when interpretable and used at first line supervision level. Too often planning is an academic tool orientated towards progress measurement and reporting and not towards aiding the effective organisation of work. Project planning should be construction driven. It should have the flexibility to cope readily with the change from area orientation at the beginning to systems orientation in the run up to testing and commissioning.
- ▲ **Safety.** An unsafe and untidy site is often an unproductive site and can usually be attributed to poorly trained and unmotivated management.

DESIGN CONSTRAINTS

Inadequate design is probably the major and most underrated factor in slowing down construction and causing productivity to fall. This could be because design decisions and delays have long been forgotten by the time their effects are felt on site. Some of the factors that need to be considered are inappropriate or over design which can be unnecessarily expensive by requiring the application of tighter standards or tolerances than are required for the end product. Design for construction, otherwise known as constructability or buildability is too rarely taken into account by designers with a lack of practical experience. This shortcoming can be overcome by assigning the project construction engineer to head office during the design phase to advise where changes to design can provide considerable savings in construction.

The effects of late design can be pernicious with the delays of one design engineer being multiplied 10 times when the delay transfers to site. Unfortunately they can be all too easily disguised by the issue of drawings on time but with 'holds' that make the information contained of value only for planning purposes. Incomplete drawings can be more difficult to finalise than if they had never been issued in the first place. 'Fast-track' design can cover up the late, or incomplete, issue of drawings on the basis that it represents beneficial overlap of design and construction.

Delays by clients in the approval of drawings and specifications fall into the same category as late design. They provide design teams with the opportunity to excuse themselves at the expense of the client.

MATERIALS CONSTRAINTS

The effect of materials problems and their impact on manufacturing processes are well enough known for techniques such as JIT (Just In Time) to be developed to overcome them. Likewise major manufacturers negotiate term supply contracts with tested suppliers to ensure the supply of material on a reliable and preferred basis. Construction materials procurement still tends to be relatively unsophisticated with at worst the purchase of some key items being left for site to buy on an 'as required' basis. Even with advanced ordering, sophisticated expediting and inspection, construct work still suffers from late delivery, incorrect fabrication and incorrect materials. Quite often these result from a purchasing policy which dictates cheapest price as the sole criteria for placing an order in preference to delivery on time, and reliability of quality. A related problem is that of out of sequence deliveries where, though a delivery date is apparently met, the materials supplied are not usable (typically, fabricated pipework deliveries comprising straight lengths but not the interconnecting 'spaghetti'). This can result from inappropriate terms of payment leading a supplier to deliver some materials quickly in order to generate income.

MINIMISING SITE WORK

Reducing the hours spent on site should be the main target of construction managers. Key areas are reducing unproductive and lost hours, improving access and constructability and removing work from site altogether.

The most radical application to minimising site work is through the construction of complete floating barge mounted process plants. At an intermediate level pre-assembly and modularisation techniques can be applied with success in providing a significant reduction in site hours. This level of application is becoming more common in the building industry, and can have its uses in civil engineering.

Pre-assembly and Modularisation

Pre-assembly usually means putting together on the ground a number of components with the objective of lifting or moving them into position in one piece. By this means it is possible to reduce inclement weather exposure, the climbing time, crane hire and associated access scaffolding for work which would otherwise be carried out hundreds of feet in the air. It also improves safety. Typically this might comprise a process plant 'column' decked out externally with platforms, ladders pipework, insulation, electrical and instrumentation systems. On a simpler scale it could comprise the pre-assembly of a number of pieces of structural steel into one section to be lifted in one piece.

Modularisation is a much more complex operation. In the process industry sections of a plant, comprising a number of vessels, pumps etc are assembled within a structural steel frame, interconnected with pipework and tested in a module fabrication yard distant from the site. The modularised units are landed adjacent to each other leaving only final closing welds to be made in pipework and the pulling through and terminating of large cables. It could comprise completely fitted out process plant control buildings or alternatively toilet rooms for incorporation in office buildings, ready to be connected to plant systems.

Reasons for Modularisation

Whilst pre-assembly has become generally accepted as an economic method of reducing construction costs, the benefits of modularisation are less easily proved. Engineering and materials (structural support) costs are higher. Combined module yard and site hook-up costs are also probably higher. The reasons for using modular construction (which will rarely exceed 50 per cent of all conventional work content) will come from among the following:

- ▲ Very confined space within an existing building or operating process unit into which the new works are to be built;
- ▲ Limited material lay down space available for conventional construction.
- ▲ Limited number of personnel employable on the site for reasons of non-availability, a severe shortage of local living accommodation, environmental planning restrictions etc.
- ▲ Severe local inclement weather conditions.
- ▲ Short project period.

Most of these reasons are those which dictated the use of modular construction techniques for offshore oil and gas production facilities as particularly exemplified by conditions in the North Sea. Other factors to be taken into consideration — though these are bonuses rather than essentials — are better quality control, better labour productivity and fewer interfaces between civil and mechanical work.

Successful Implementation of Modular Construction

The successful implementation of modular construction concepts requires radical rethinking of a contractor's procedures, systems and programmes. The following factors need to be taken into account:

- ▲ Everyone associated with the project needs to think in terms of modules.
- ▲ Module sizing relates to transportation limitations.
- ▲ Specification of fireproofing, insulation and painting has to be established out of normal sequence at the commencement of the project.
- ▲ Electrical and instrumentation design engineers need to plan and rethink their ideas on the sequencing of their activities and of the design of systems within modules.
- ▲ The capability of module yard vendors in respect of quality control and delivery on time is crucial.
- ▲ Prepurchasing of materials for free issue to the module yards. Programme restraints will not usually permit supply by the yard operators. This will require acceptance by the main contractor of commercial responsibility for quality and delivery on time.

- ▲ Control of supply of materials to module yards and sites is critical and may require the establishment of an intermediate warehouse.
- ▲ Consistency in application of modular construction principles to all aspects of the project (eg package units such as air coolers being provided completely prepackaged through to platforms and lighting).
- ▲ Interconnection of module pipework can be by utilisation of 'pup' pieces or by exact matching. The latter emphasises the need for the application and control of tolerances to all aspects of work, from foundations through structures to setting equipment, and fabricating and erecting pipework, to a degree previously not considered practical.
- ▲ A consistent QA/QC plan for the complete project.

Transportation

Transport of massive modules to onshore locations and particularly within existing operating process units requires detailed pre-planning. It may require the removal of much street furniture and some roads and bridges may need widening or strengthening (modern units should be designed to cater for the movement of large components). A plan of action should be drawn up to cater for such emergencies as blocking a road normally used for emergency services vehicles.

Industrial Relations

The rapid and successive delivery of a number of modules to site will quickly open a number of workfaces. This will result in a rapid increase in the numbers of men employed on site. This in turn requires pre-screening of potential employees and a rapid turnaround in recruitment and induction to take advantage of the work available. If not properly handled in conjunction with the trade unions involved it is a source of potential problems.

Safety

A benefit of modular construction is the improvement in the management of safety. It minimises the risks of men working at heights; it minimises the number of scaffolds which have to be erected (work can be carried out from the permanent platforms); there is less construction debris generated and therefore fewer underfoot hazards; permanent lighting is available relatively early.

Misconceptions

Common misconceptions in the application of modular construction are:

- ▲ Design work must be frozen before module fabrication and construction starts.
- ▲ 95 per cent material definition is required before orders are placed.
- ▲ 80 per cent engineering definition required for enquiry issue.
- ▲ Quality control of a higher standard than normal is required.
- ▲ Higher trade skill levels than normal are required in module yards and on-site.
- ▲ Full time trades union officials need to be involved in the project from the outset.
- ▲ Planning needs to be construction driven.
- ▲ Construction plot must be paved before mechanical work can be started.

Identified as misconceptions, these widely quoted essential requirements for modular construction are in fact no less needed for the success of conventionally constructed plants. Prior to the inception of work in the North Sea it was commonly perceived by clients, project managers and design engineers that design changes and material delivery delays could readily be catered for within the construction programme float. The pressure of huge investments at stake in the North Seas combined with imperative North Sea weather windows caused a radical rethink of front-end requirements. However in some circles these requirements are still seen as peculiar to modular construction.

IMPACT OF SITE LAYOUT

Badly laid out sites can considerably increase the cost of construction.

Permanent works are usually laid out with two criteria in mind. One follows the flow of materials from the raw state to the finished product. The second groups equipment together to minimise the costs of interconnecting materials. Very often these criteria lead to layouts which made access both for men and machines difficult and expensive. Costing construction work is still an inexact science. It is not possible therefore to calculate the savings resulting from a more generous spacing of plant. Construction engineers will always delight in solving some designed-in difficulty. It is however more realistic to design them out in the first place.

Poorly thought out site access and internal access roads can be expensive when the effect of width restrictions and busy junctions leads to traffic queues at the entrance to and within the site. Holding up the movement of cranes, plant and men can quickly lead to delays and increased costs. Traffic lights and banksmen controlling traffic at junctions are an indication of poorly thought out site access problems.

The use of large cranes has reduced the cost of labour previously spent erecting gin poles and guy derricks. However, careful thought still needs to be given to access for cranes in the layout of site and permanent roads within process plants.

Catering for long delivery items should be taken care of by locating them at the periphery of sites where their delivery at a late stage will cause minimum disruption.

OVERTIME AND SHIFT WORKING

When programmes start to slip one commonly accepted solution is to increase the hours worked on site. Where a site has reached saturation point this might indeed be the only solution. However have you considered:

The effect of overtime deterioration. The result of tiredness and boredom after working 60 hours a week for a long period, will soon wipe out any benefit (see Fig 12.3 and [Ref. 1]).

Absenteeism will increase and make the problem of balancing gangs even more difficult. The resulting additional hours paid at premium rates then become even more unproductive and expensive.

The effect you are trying to achieve may be better achieved by selective targeted overtime for the particular section of work that is causing the bottleneck, or by bringing in a separate night shift crew (probably a subcontractor) to clear up the backlog.

The alternative of shift working is generally a better solution. However, it needs to be built-in as part of the original project philosophy. Attempting to negotiate it at a late stage will only lead to resistance from the workforce. It will be seen as speeding up the time when they will be put out of work without the benefits they might otherwise obtain from overtime payments.

A tight programme or a particularly compactly designed plant with a low labour saturation level may make it necessary to build the plant from the outset under a shift work agreement.

Figure 12.3 *Deterioration in Performance with Increasing Numbers of Weeks of Scheduled Overtime Working*

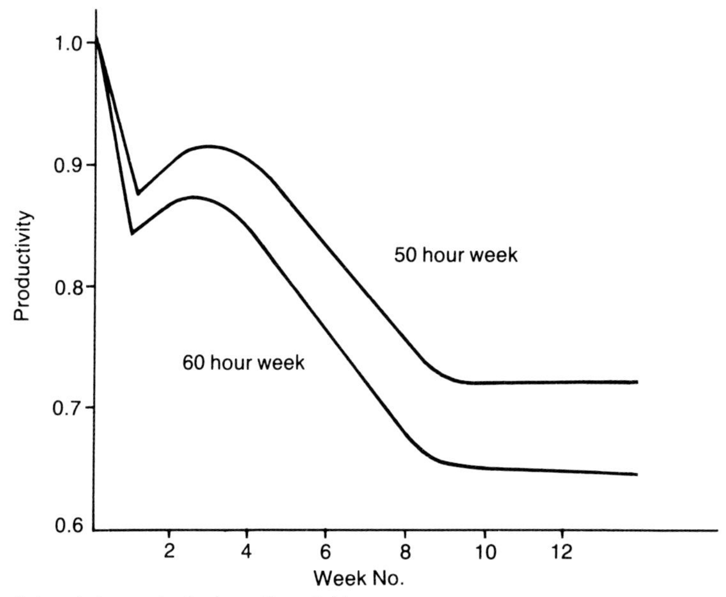

Acknowledgements: Business Roundtable

There are however the benefits of the alternative of rolling shifts. With rolling shifts separate construction crews are employed to work longer hours over consecutive periods (typically four days). The problems of overtime deterioration are avoided as the non working crews are being adequately rested. It may be particularly effective for a plant in a comparatively remote location. It does, however, require additional management staff and supervision.

Shift working can reduce the construction period, by making the best use of the working day and the high capital value of construction plant.

The problems of shift hand over are resolvable and may be offset by the competitive spirit that can be generated and by reducing the problems of unbalanced gangs.

FINISHING CONSTRUCTION

Identifying the cut off point of construction and hand over varies between clients and reflects to some degree the responsibility that individual operators want to take for, or participate in, pre-commissioning and commissioning. Once commissioning starts there may be a disruptive effect on construction resulting from restricted access to working on a live electrical systems or under hot-work permit systems. This needs to be recognised and taken into account in the planning and costing as well as organisation of the final stages of a project.

Due to difficulties in measuring completed work any attempt to speed up completion by the payment of site wide completion bonuses, by gang bonuses (job-and-knock) to complete individual sections or work are doomed to failure. They have not only been shown to be ineffective but are divisive and capable of causing industry wide disruption. They do not work. Do not be tempted into paying them, even for the supposed 'final' month.

REFERENCES

1. *Scheduled overtime effect on construction projects* — The Business Roundtable November 1980
2. *NEDC Large Sites Report* 1970

13 IMPROVEMENT AND MEASUREMENT OF PRODUCTIVITY AND PROGRESS

IVOR WILLIAMS
National Economic Development Office

PRODUCTIVITY DEFINED

Productivity is defined as the ratio of output to input. It results from the efficiency of the productive process, in particular the efficiency with which labour and capital (men and machines), are used in transforming inert materials into socially useable end-products.

Achieving greater efficiency and productivity is important not only to the individual project or undertaking, but also to the industry and to the nation, since in a highly competitive world only the efficient survive. This is well recognised by the Japanese. The mission statement of their National Productivity Centre includes the following:

> 'Productivity is, above all else, an attitude of mind. It is a mentality of progress, of constant improvement of that which exists. It is the certainty of being able to do better today than yesterday and less well than tomorrow. It is the will to improve on the present situation, no matter how good it may seem, no matter how good it may really be. It is the constant adaptation of economic and social life to changing conditions; it is the continual effort to apply new techniques to new methods; it is faith in human progress.'

THE DIFFICULTIES OF PRODUCTIVITY IN THE CONSTRUCTION INDUSTRY

The optimisation of productivity, whatever the product, requires that all aspects of the production process are scrutinised to reduce unnecessary time and effort. Whether it be for the manufacture of widgets or the execution of a construction project, production is an integrated process comprising many parts, and only a systematic analysis of the entire process will yield significant results.

In the construction industry this analysis must include both the on-site work and the off-site activities of design, planning, procurement and delivery, when many of the conditions for high productivity on the project are determined. However hard men may work on-site, their output may be affected by problems not foreseen during the design and planning stage, such as the unavailability of access to the workplace or the remedying of errors due to incorrect drawings.

The construction process differs significantly from manufacturing in that many construction projects are unique either in terms of design or site location. Each one involves a new learning process and on each one new problems will emerge. Even where a project is repeated, its sheer duration generally means that the project team, both management and workforce will have undergone considerable turnover, militating against the more efficient operation that repetition and familiarity ought to produce.

There are other circumstances, external to the project and largely outside the control of the project team, which may adversely affect productivity. The casual and short-term nature of the employment relationship which is normal to the industry may undermine the commitment of men. Historically, the training of operatives within the industry has been inadequate. Few people have been trained specifically by and for the industry. Even where this is the case, much training tends to accord to rigid craft divisions, is general and is not specific to the needs of the industry. Additionally, although each project, whether from an employer or union perspective, is responsible for the harmony of its own industrial relations it will not be unaffected by industrial relations on other construction projects or within the economy generally. Such problems are best tackled on an industry-wide basis, but despite all these constraints, a great deal can be done to optimise productivity on any given project.

Optimising the productivity of a project can be approached on two fronts. First on the organisational and planning front, the conditions for optimal productivity must be considered and met in the planning and design stages. The second front is that of motivation. Once the organisational and logistical preconditions for success have been taken into account, the entire project team, including the management and the workforce, have to be encouraged and motivated. The responsibility for productivity optimisation ultimately rests with the project manager. Only he has the remit which embraces the entire process from conception to commissioning.

THE ORGANISATIONAL ASPECTS OF PRODUCTIVITY OPTIMISATION

Productivity optimisation begins with the planning for the execution of the project. The project master plan details all the activities necessary for the completion of the project, their logical sequencing and the essential control mechanisms. It is the fundamental tool which enables the project manager to steer his project according to its budget and programme objectives and is the bedrock upon which productivity optimisation stands.

Productivity optimisation must be considered as early as the design phase if it is to be achieved over the entire project. Design has two aspects. First, there is the process or conceptual design which aims at ensuring that the completed project meets the client's requirements and operates efficiently. Secondly, the engineering design is concerned with finding engineering solutions to the requirements of the concept or process, but these solutions must also facilitate construction. Such consideration will only receive sufficient attention where the same contractor is responsible for both design and construction.

The term for these considerations is 'constructability', 'buildability' or 'design for construction'. They should include such matters as sequencing and accessibility and also whether the project should be constructed in the traditional way or preassembled and modularised. The latter method recognises that the fabrication and assembly of large parts of the project might be more efficiently carried out under workshop conditions than on the site. It is a process which is becoming ever more popular in the building industry and is now successfully carried out in the process industry with modules weighing up to 5,000 tonnes. An additional advantage is that it allows assembly off-site to run in parallel with civil engineering work on-site, but for units which are larger, the availability of access both to the site and on the site is vital. Modularisation often leads to heavier structures as additional steel members are required for stiffening purposes during transportation and installation.

The design process has traditionally been very labour intensive. Computer Aided Design (CAD) can transform this process by significantly reducing the number of draughtsmen required although its full potential in this respect has not yet been realised. CAD should reduce errors, particularly of access and tracking which can be more easily spotted in the three-dimensional mode available on computers. The rectification of errors during construction is expensive (often 10 times the cost of rectification during design), affects schedules, and undermines morale, the maintenance of which is also essential to good productivity.

A further requirement of the design process is that it must not delay the early ordering of materials and equipment, any such delay might affect schedules on-site. This consideration would not be appropriate where the technique of 'fast track' construction is being employed. By increasing the overlap between design and construction, fast track has the potential to shorten the programme significantly, and has been used with great success on building and civil engineering projects. It has less relevance to more complex engineering construction projects where experience has shown that each of the preceeding stages of design and procurement should be substantially complete before work begins on-site.

The timely procurement and delivery of materials is a critical part of the project process. To enable progress on-site to continue uninterrupted, orders have to be placed in sufficient time and with suppliers who are able to meet the delivery requirements of those working on site. 'Control' passes temporarily from the project team to a manufacturer or fabricator, and greater control may be retained if a resource controller is appointed; who will maintain close liaison with the supplier, helping to avoid the manhour cost of 'lost' time which may be caused by delivery delays.

Delivery itself has to be carefully planned so that the equipment can be sorted in a manner that will allow easy access and retrieval as and when required. The 'Just In Time' technique, in which deliveries can be taken direct from lorry or barge and erected without storing, and therefore double handling, can save on-site manhours, but on complex projects is almost impossible to co-ordinate. Materials and equipment on-site, accessible and waiting to be installed, are probably the best insurance policy.

These are the necessary preconditions for successful and efficient construction of the project. They are not, however, sufficient in themselves. Loss of control during the labour-intensive and heavily interdependent construction phase can render even the most thorough planning useless. The layout and infrastructure of the site itself requires careful consideration to eliminate (as far as is possible) bottlenecks affecting the movement of men, machines and equipment, which cause wasteful lost time. Consideration must also be given to optimising the working hours and establishing the most productive arrangement of the working week. Shift work should be considered, but overtime on a scheduled basis should be avoided. Critical attention should also be given to the size and composition of the labour force; productivity can often be improved by cutting back the number of men on-site.

Lost time is probably the greatest detractor from the achievement of good productivity during the construction stage. It includes time lost through disputes and unauthorised absenteeism; it must also include all time when men are not engaged in productive work. The causes of this extra lost time are walking time between the workface and the amenities, ie toilets, cabins, canteens, etc; and the late starts and early finishes in the mornings and evenings and time lost around tea and meal breaks. It should also include the time when men are unable to work for whatever reason, ie waiting for materials, drawings, access, or for the assistance of other craftsmen with whom their work interfaces. Control of lost time is essential to good productivity since experience shows that a considerable proportion of potentially productive time is wasted.

Activity sampling,[1] a technique used by the National Economic

---

[1] The techniques of Activity Sampling and Foreman Delay Surveys are explained in a NEDC publication 'Promoting Productivity in Construction', available from NEDO Books.

Development Office (NEDO) in many comparative studies, has shown that in fact it is not unusual in the construction industry for only around 20 to 30 per cent of the day to be spent productively. There is massive scope for improvement here. Another technique, the Foreman Delay Survey can assist in identifying the causes of lost time. Fundamentally, however, the problem relates to work planning and work co-ordination, motivation and the need for effective supervision.

Work study techniques have played virtually no part in the construction industry in the past. This has been due to actual or potential opposition from the trade unions or workforce. The situation has however changed and the importance of good productivity is recognised by all. There is no reason why, if a total project approach is to be taken to productivity optimisation, as this chapter suggests, the techniques referred to above and the technique of Method Study cannot be more freely introduced and workforce participation invited.

Motivation[2] is a particular problem in the construction industry due to the casual nature of employment. A great majority of the men, including many in managerial and supervisory positions, can expect employment only for the duration of the project and can expect to be made redundant at completion. They have little reason for commitment either to the project or their employer and have no real interest in working more productively.

Attempts to build up commitment take two forms: one is the positive motivator of monetary incentives through various schemes, productivity bonuses, end-of-job bonuses, target payments, etc; the second is the removal of demotivators. Monetary incentives can work effectively only if the operative or recipient can relate his pay directly to his own effort. This may be the case in the building industry, but in engineering construction is rarely possible, so resort is made to group bonus schemes.

The integrity of such schemes is often open to question; more importantly the incentive effect of group bonuses is low and there is no evidence to suggest that they actually promote productivity. Furthermore, bonus schemes have historically been the most frequent cause of dispute on-site and hence become a significant demotivator with consequent detrimental effects on productivity.[3]

Other means of building up commitment, attempt to create a sense of identity and pride in a project. They include Project Joint Councils, briefing circles, the provision of site newspapers and social activities,

---

[2] A great deal of work has been done on the motivational aspects of productivity by Dr John Borcherding of the University of Texas.

[3] The author recognises that there may be other points of view on this matter and acknowledges that specific incentives to meet specific objectives may be effective, but believes that such expedients must be balanced against the longer-term need for uniform payments systems.

family open days, inter-work group competitions of various kinds and other initiatives. All are important, as is good communication generally.

The other approach is the removal of possible demotivators. These include ensuring that due regard is paid to all on-site conditions such as canteens, and changing facilities; paying due regard to the men's problems; the speedy resolution of disputes; treating men and their unions with respect; and establishing good industrial relations generally. All can be of assistance. The greatest demotivator, however, is the impermanence of employment and the provision of greater continuity would make the most significant contribution to maximising productivity.

Good supervision is essential. The supervisor is at the pivotal point between the planning and design intentions and their achievement on site. His responsibility is for output, the input factors having largely been determined. He more than anyone else is responsible for the reduction of lost time and for ensuring that all possible constraints on output are removed. He must establish a strong rapport with the men under his supervision, but must be fair and authoritative in his single-minded pursuit of output. His qualities, among which leadership is paramount, should be exceptional and he should be rewarded in salary, status and security. The reality in the construction industry falls far short of this ideal. He is usually insecure, inadequately trained, poorly paid, given little authority and is invariably selected on criteria other than his leadership qualities. Consequently his loyalties are divided and productivity suffers.

PRODUCTIVITY AND PROGRESS MEASUREMENT

The measurement of productivity is inextricably bound up with, and depends on, the accurate measurement of progress. Productivity was defined at the start of this chapter as the ratio of input to output. The input in virtually all industries is measured in manhours. Though production requires both labour and capital, these are reduced to a manhour unit which recognises the enormous importance of labour, particularly in the construction industry, and provides the standard unit that is required.

Output is measured according to the work being undertaken, whether it be the laying of concrete or the erection of steel. Thus, for example, productivity can be indicated by the number of manhours required to lay a cubic metre of concrete or to erect a tonne of steel. This however is not a true ratio as the units are not common.

Output can be expressed as manhour units through the medium of 'norms' or estimating standards. These are estimates of the manhours required to complete a given amount of work, ie erect a tonne of steel etc. Thus to establish productivity as a ratio, the manhours estimated to be necessary to complete the work (output) is divided by the actual manhours required for its completion (input). The ratio or productivity factor will be less than or greater than unity according to whether the actual manhours were greater or lesser than the estimate. A figure greater than unity indicates productivity better than estimated.

The reciprocal of the productivity factor is the cost factor. Thus where, as in the above example, the actual manhours are less than those estimated, the cost is also less than anticipated.

Norms can be established by work study methods or by drawing on experience. However each major contractor will have his own data bank of norms which will assist him to estimate his contract bid. Measuring progress against his original estimate is an essential indicator of the well-being of his contract.

To be meaningful, both input and output have to be accurately measured. On the input side, exactly whose manhours should be included in the equation, ie direct and indirect operatives, has to be determined. On the output side, the work must be broken down into small but realistic and relevant detail, and the manhour content established by application of the norms. By this means it is possible, simply and with some accuracy, to measure the progress of work within each activity. The summation of work within each activity can then be carried out across the entire construction network and progress on the project for the period under consideration accurately measured.[4]

For accurate measurement of progress throughout the entire period of the project, accurate estimates of the original work scope manhours are an essential prerequisite. The second requirement is that the work scope should be accurately, quickly and regularly updated to reflect any changes made. The third requirement is that the volume of work remaining to be completed is regularly updated to reflect the progress and productivity achieved and that which is required for timely completion of the project.

The measurement of progress is often a contentious point between the client and his contractor, since it may have commercial implications if progress payments have been agreed. For this reason, and also because it is fundamentally important to project control, it is desirable that it should be carried out on a regular basis by independent surveyors who are able to report impartially to both parties.

Progress is usually plotted on what have become known (in the UK at least) as 'S' curves. The shape of the curves (Figure 13.1) reflects the slow start up and 'learning curve' process in the early stages of the project, after which progress accelerates and is reflected in the steeper slope of the curve. Finally the curve flattens out, reflecting what is known as the 'five per cent problem', where the final five per cent of the work, which

---

[4] For further information regarding this system of progress or output measurement, see Chapter 19, Project Planning and Control by A. Lester, Butterworths.

incorporates the minutiae of the construction process, takes perhaps 15 per cent or more of the project duration. The flattening may also reflect the fact that for much of the workforce, redundancy draws near. Progress curves of this shape are by no means inevitable, and project teams should aim instead for curves such as the one in Figure 13.2 where after a minimal build-up period, a more or less constant rate of progress is achieved throughout the project. (Both curves are from actual projects.)

Figure 13.1

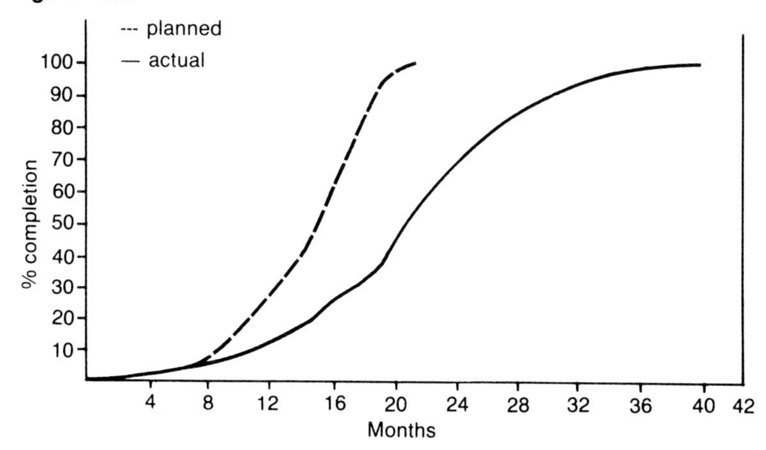

Illustration to demonstrate actual against planned progress on a large chemical plant project completed in the 1970s. Apart from the considerable lateness of completion, the familiar S shape of the progress curve shows the slow beginning and an extended completion period.

Figure 13.2

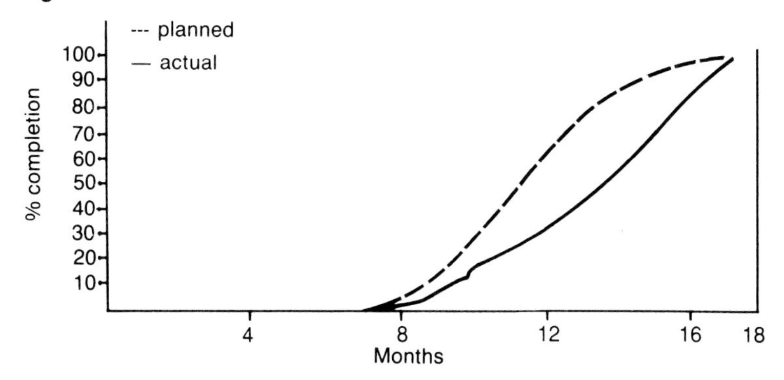

This illustration demonstrates an on-time completion of a smaller chemical project, but note how after a relatively slow start-up, progress accelerates and maintains an upward rate to the end of the project.

INDUSTRY-WIDE PRODUCTIVITY

It has been suggested earlier that some of the preconditions for achieving good productivity on construction projects are laid down at industry level. These include appropriate and satisfactory training; the industrial relations framework; and possibly means of retaining skilled and valued labour through a registration scheme which will at the same time reduce some of the insecure and short-term employment factors which are so costly to productivity. Productivity should also be measured on an industry-wide basis. Whether the industry is improving or otherwise remains a matter of subjective opinion.

Fundamental to industry measurement is a single data bank of norms appropriate to all manner of work undertaken by the industry. Such a data bank has been produced by NEDO and remains available, but so far has not been adopted by the industry. More recently NEDO has proposed a Standard Productivity Measurement System which reduces all work to 10 categories and produces standard norms for these categories. Adoption of the NEDO system will allow measurement between projects of different types carried out by different contractors and will indicate whether efforts to improve productivity are being effective.

14 INDUSTRIAL RELATIONS

SIR PAT LOWRY
National Joint Council for the Engineering Construction Industry

It is a truism that good industrial relations are vital for two reasons — first to enable management to achieve its business objectives, eg the profitable completion of a construction contract on time to the requirements of the client; and second, to provide employees with better job security and improved terms and conditions of employment. Good industrial relations do not just happen; as in every other area of management activity, they require dedicated forward planning and meticulous attention to detail.

Maintaining good industrial relations is not just the responsibility of the industrial relations manager. Every manager must accept that it is his task to lead and motivate those under his command, to ensure that grievances are promptly handled; and that essential disciplines, including those relating to health and safety, are observed.

This does not mean that every manager has complete freedom to operate as he sees fit. The pay and conditions of employment of the workforce as a whole may be determined by higher management, perhaps as a result of negotiation with recognised trade unions. The negotiations may indeed be carried out on a multi-employer basis leading to a national agreement between an employers' association and a trade union or a number of trade unions. Every employer who is a member of that employers' association is then under an obligation to observe its terms.

It is fashionable today to criticise multi-employer agreements on the grounds that they are inflexible, and that they place unnecessary constraints upon employers individually to pay whatever wages and to apply whatever conditions of employment the business can afford. In some industries, tightly-controlled multi-employer bargaining has certainly outlived its usefulness: elsewhere (and the engineering construction industry is as good an example as any) there are sufficiently sound historical reasons for keeping it in being. One of the key aspects of the forward planning to which I have already referred is the need to keep under constant review the nature of the agreement which is likely to serve the needs of the industry best.

Collective agreements, whether applicable to a single employer or to a whole industry, contain a number of common elements, eg rates of pay, conditions of employment and a procedure for handling claims and grievances. But if there is one issue more than any other that is likely to cause disputes between an employer and his workforce it is the way in which productivity and efficiency are to be measured and rewarded. Some employers prefer to pay their employees on a fixed rate basis and to depend on their managerial competence in maintaining the required level of production. Employees are paid their share of any improvement at those times when the agreement is open for renegotiation. Other employers, perhaps under pressure from their employees, prefer to reward productivity improvements more immediately, through an incentive bonus arrangement.

If serious disputes and a consequential inability to meet the commitments to the client/customer are to be avoided, it is vital that there are clear provisions as to how output is to be measured and improvements in output rewarded.

However clear the terms of an agreement and the provisions of any supplementary rules prescribed by the employer (eg in relation to timekeeping and discipline), it is almost inevitable that from time to time an employee will feel that he has been harshly treated. To ensure that these grievances are promptly and fairly handled the collective agreement will require a grievance procedure of which the essential elements are:

- ▲ A number of stages which enable unsolved grievances to be promptly handled at higher levels within the managerial and trade union structure.
- ▲ Prescribed time limits within which each stage of the procedure must operate.
- ▲ Provisions which clearly specify the obligations imposed upon the employer before he is free unilaterally to vary terms and conditions of employment.
- ▲ The peace clause under which industrial action of any kind is ruled out until the grievance procedure has been fully carried through.

When all negotiations have failed, some grievance procedures permit the registration of an inconclusive failure to agree. But others are of a kind which at the end of the road require a decision to be reached which both sides are then under an obligation to accept (the engineering construction industry is a case point). Procedures of the first type open the way to industrial action once a failure to agree has been recorded at the final stage; procedures of the second type produce decisions which avoid a failure to agree and thereby make industrial action unnecessary. Nobody calls them no strike agreements but they are certainly 'no need to strike' agreements.

In negotiating the grievance clauses in their agreements, employers and trade unions must be very clear as to what kind of procedure they prefer.

Employee grievances, if they are not to fester, require prompt handling. More than that they demand managerial firmness (not to be confused with inflexibility), fairness and consistency. Of these three qualities it is perhaps consistency that is the most important. The management that reacts to an unofficial employee walk-out by refusing to negotiate on day one, but subsequently engages in 'talks about talks' or makes positive concessions while the strike continues, does little or nothing either to preserve the credibility of the negotiating procedures or to discourage further walk-outs. But the management that always reacts to an unofficial employee walk-out by refusing to negotiate until work has been resumed is likely in the long run to enjoy more peaceful industrial relations. The management that handles disciplinary issues on the basis that the same offence is likely to incur the same penalty is more likely to secure compliance with essential disciplines than the management that blows hot and cold.

Prompt and consistent grievance handling is therefore essential, but far more than this is required if management is to secure a high level of employee commitment. A vital element is good communication. Employees are entitled as a matter of right (not of managerial whim) to have answers to the following questions: What do you want me to do? How am I doing? What are my prospects for the future?

Increasingly, too, employees expect to be told how the work they are doing fits into the wider framework of the employer's activities. How is the current contract proceeding? What are the problems? What does the future order book look like?

It must never be forgotten that the one person who can make the most constructive suggestions as to how the job might be carried out more effectively is the employee himself. This is just one reason why employee communication is a two-way process. Managers must certainly communicate but they must also be prepared to listen; and they must do it regularly. Good communication cannot be casual or irregular. It must be methodical and organised to a prescribed system so that all managers right down the level of first-line supervision know what is expected of them on a regular basis.

The ability to lead, to motivate and to communicate is so vital today that technical competence on the part of managers, important though it is, is no longer enough. Increasingly their training must include human resources management as well. Indeed their initial selection as managers increasingly requires this vital aspect to be taken into account in the recruitment process.

Employers thus have the responsibility to communicate about those matters on which they wish to keep their employees fully informed. This is not a trade union responsibility. But a good management communication system and trade union recognition are not incompatible. There is no law today which compels or pressurises an employer to recognise a trade union. But where it is clearly the wish of the majority of the workforce to be collectively represented, he would be unwise to decline. Where trade unions are recognised, they are entitled to reasonable facilities to enable them to report back to their members on the outcome of negotiations with the management. They must also carry their share of the responsibility for ensuring that their members comply with the agreements negotiated on their behalf, including in particular the 'peace clause' in the grievance procedure.

And just as shop stewards or site stewards are entitled to reasonable 'report back' facilities, so too is the workforce entitled to decent facilities in which to prepare for work, to eat and to attend to the needs of nature.

Working life on a construction site will never be a bed of roses but if workers are to give of their best, they are entitled to a management which caters properly for their creature comforts. It is all part of the motivation process. The acceptance of filthy conditions at the place of work almost inevitably leads not just to shoddy work but to poor discipline as well.

Industrial relations in the UK are currently in a state of flux. New ideas abound. We must always analyse these carefully and determine their relevance to our work situation. But, as much as anything, improvements in industrial relations will depend on better planning and our ability to carry out just a little better all those things on which we have traditionally concentrated. We must continue to aim for better trained managers, clearer and more comprehensive agreements, more effective grievance procedures, better employee communication and improved 'welfare' provisions.

15 COMMISSIONING

TREVOR LANE
Foster Wheeler Energy Limited

In process plant construction, the commissioning phase has always been recognised as a vital part of the project. However, with the increasing sophistication of services in buildings in recent years, this sector of the construction industry now needs to adopt a similarly rigorous approach to this aspect of the work.

This chapter is intended to address the commissioning phase of both types of construction projects.

Commissioning is a process which can take a matter of weeks on a well designed, well constructed plant, but many months on one poorly designed and constructed eating substantially into the end user's financial return.

Commissioning personnel can themselves only marginally influence the length of time the activity takes. However, the earlier the commissioning manager and senior staff are involved in the project, the less likely it is that their influence will be adverse. It is a false economy to involve commissioning personnel only when construction work is 90 per cent complete.

Professional commissioning personnel, in the absence of experienced client operators, have much to offer during the process design phase in identifying potential operation economics, and their advice should be obtained at this stage.

In order to achieve the objective of a rapid start-up certain actions and procedures must be implemented long before commissioning starts. This chapter contains salient guidelines for a commissioning philosophy. They need to be expanded to suit individual projects.

WORK SCOPE AND PRIORITIES

Long before construction is finished it is necessary to establish the commissioning scope of work, ie:

- ▲ identify the fluid content of each and every system, the major items of equipment, the pipe lines, the instrument loops and so on;
- ▲ identify which systems require chemical cleaning and its extent;
- ▲ establish the extent of water flushing and steam blowing required;
- ▲ identify the valves, instruments etc that need to be removed to enable the filling of the above to be implemented;

- ▲ source and have available temporary piping, spool pieces, silencers and the like;
- ▲ prepare schedules for initial charges of lubricants and greases and define clearly the extent of systems that, after cleaning, will require to be preserved, ie nitrogen purged;
- ▲ identify and quantify all raw materials, intermediates (and where necessary final product — sometimes required for systems testing, seal filling etc) required for start up.

Normally early construction activities are carried out on an area basis. Through construction testing into precommissioning this will change to completing by systems. Common logic will usually establish the sequence in which the main systems are completed and tested, eg large bore before small bore, water systems before steam or process fluids. However, it is a key responsibility of the commissioning team to define the priorities of all other systems and sub-systems.

Ideally the commissioning 'system orientated' plan should be established early in the design phase when it can most usefully influence design and construction programmes. One particular benefit of this is that any additional isolating valves needed for discrete area commissioning can be incorporated at the appropriate early stage.

In addition to defining the work scope, the commissioning team is concerned with collecting all documentation related to plant eg purchase orders, radiographs, mill certificates, hydraulic test records etc. An effective QA system will of course ensure the orderly collection of these documents for commissioning purposes.

WHEN TO START

Experience shows that an early start on commissioning does not necessarily mean or ensure early programme completion. What is important is the proper phasing of work based on the project's start-up priorities. This will normally mean first commissioning the electrical and water systems. With these two complete, motor rotation and vibration checks can be carried out and pumps made available for system flushing. Early completion of permanent lighting not only aids construction but also the precommissioning activities. These, by their very nature, are disruptive to the construction workforce and are therefore usually carried out after construction has finished for the day.

Long duration activities such as instrument loop checking and lube oil flushing associated with large rotating equipment should be identified and programmed in at the appropriate time.

Precommissioning always starts before construction is finished. It is therefore important to recognise and implement those activities that can be carried out without unduly prejudicing the completion of construction.

PLANNING AND MANPOWER

Having defined the work scope and system priorities, the precommissioning and commissioning plan must be made and resource loaded.

This plan must be the document for driving the starts and finishes of all of the phase priority tasks required to meet the overall start-up sequence.

Resourcing the work will almost certainly involve instrumentation and control systems manufacturers' engineers on site during the precommissioning and commissioning stages, as well as those from the makers of large rotating equipment. Manufacturers must be advised in good time if they are to ensure availability of their personnel when required. Early consideration must also be given to the rigging, pipefitting, instrument and electrical trades personnel required for commissioning work in addition to construction completion. It is not realistic to expect construction personnel to be available on an unscheduled basis.

COMMUNICATIONS

It is essential that regular action meetings are held during commissioning stages of a project.

A meeting should be held at the beginning of each working day and should involve project, commissioning and construction managers. If the project is being run on shifts, meetings should be held at the beginning of each shift.

Structure the meetings. Briefly review progress achieved. More particularly search out, identify and anticipate problem areas and possible delays.

Ensure that people attending the meeting are those responsible for getting work done. The meeting is for co-ordinating all areas of work and ensuring that efforts are directed towards the critical activities. Should delays manifest themselves, action is needed either to provide more resources or to reprogramme subsequent activities.

This should be a dynamic meeting not a technical discussion. It should not be a forum for resolving technical problems, but should contain the correct mix of people to identify problems and appoint co-ordinators for them.

PERMITS AND SAFETY

When dealing with steam, nitrogen, hydrocarbons, process chemicals, etc. all commissioning activities must be actioned under a 'Permit to Work' system with appropriate procedures for obvious safety reasons.

When dealing with electric systems an 'Isolation Certificate' must also be obtained, together with the relevant work permit.

All electrical systems should be locked off prior to energising and all valves should be tagged for open/closed position. The permit system must be controlled under a central authority and must identify the persons responsible for the work.

The permit system must be established early, co-ordinating construction commissioning and, where appropriate, any existing plant procedures. The workforce must be trained to follow the system prior to its implementation. As the plant is progressively brought on stream a register must be maintained, and site notice boards displayed, to ensure that everyone is aware which parts of the plant are live.

PEOPLE IN COMMISSIONING

Because of the long hours required, commissioning places great demands on everyone involved. The commissioning manager will be on call 24 hours a day often for weeks on end. His team will often work 12 hours a day plus the time spent handing over between shifts. It is a challenging and satisfying period, providing a stimulus that largely overcomes the problem of unsocial hours.

Commissioning appears fraught with potential industrial relations problems — different working practices side by side, wage differentials, requirements for shift working, stringent safety requirements and the prospect of imminent lay-offs. The fact that disputes rarely occur is probably due to the morale and momentum developed in this phase of the project, to the awareness by the workforce that safety is paramount, and to the extra effort which goes into communications.

BUILDING SERVICES

In recent years major changes have occurred in the design of new buildings. The requirement to ensure that the internal working environment remains constant in terms of lighting, temperature, humidity and air flow even though the outside conditions are varying by the minute, together with the need for complex communications and information transfer systems, have put a high priority on building services. Building services like other facilities have therefore become more complex.

TIMING/PROGRAMME

Commissioning must commence early. Indeed commissioning must be considered as early as the design stage. Immediately a contractor is appointed and the sequence of construction/completion agreed the following should be considered:

▲ parameters for commissioning should be set. For example if phased construction is to be adopted, consideration should be given to ensure that all systems are compartmentalised and can be commissioned separately and in the same sequence as the phased construction;

▲ a detailed commissioning programme should be produced in conjunction with the main contract finishing requirements;

▲ a detailed method statement should be produced;

▲ agreement on the specialist personnel who will be responsible for carrying out commissioning operation should be reached. Consideration should be given to employing independent specialists.

TESTING

The success in achieving completion of the commissioning of any project is initially dependent on the careful monitoring of the installation as the works proceed.

Upon completion of manufacture, or before despatch in the case of stock items, all materials, apparatus and equipment to be incorporated should be tested.

The results of each and every test carried out should be accurately and comprehensively recorded on a form and a test certificate signed by the person in charge.

Every test certificate should include full details of the item including reference numbers, the date and time of the test, the ambient conditions, a fully detailed description of the tests carried out, the results obtained and any relevant performance curves.

All test certificates should be collated on-site and related to the location of the installed item.

As the majority of any service installation is usually hidden by either the building fabric or the internal finishes, checks should be made to ensure that the correct equipment is being installed in the correct position and that each item is in working order.

This will ensure that:

▲ any problems can be resolved at an early stage;

▲ standards and quality are maintained;

▲ the need to remove finishes to gain access and the disrupting affect this has on other trades is avoided;

▲ works are completed cost effectively.

Similar test certificates/reports to those produced for off-site testing should be completed for on-site checks. Again these must be collated with the previous records in order that complete records are available.

PRECOMMISSIONING

Following the continual process of checking and testing throughout the installation period, precommissioning must take place. This phase of work takes into account the activities necessary to advance the installation from static completion to the commissioning phase.

At this time the permanent power, fuel and water supplies must be available.

Electrical Installation

During a pre-commissioning phase the following tests should be carried out:

1. Cable Testing

* Insulation resistance test
* Voltage withstand test
* Earth continuity test
* Phase rotation and phase-correspondence

2. Low voltage switchboard tests
3. System and equipment earthing

Mechanical Installation

During precommissioning the following tests should be carried out:

1. All hydraulic systems should be thoroughly flushed to remove any residual matter within the pipework systems.
2. All ventilation systems should be cleared of any obstructions and debris.

3. All fire dampers should be checked in the open position.

4. All wiring in control panels should be checked for loose wires.

5. All terminations to all control equipment should be checked against the wiring diagrams.

Public Health

Precommissioning checks on the public health installation should include the following:

1. All underslab drain runs are cleared of all debris.

2. Manholes are complete including step irons and correct type of cover.

COMMISSIONING

Commissioning unlike testing and precommissioning can only be carried out when the other construction operations are virtually completed. It will normally be essential that items such as suspended ceilings, partitions and doors are all completed. The building should also be clean as dust in the air can affect sensitive sensors. Equally it is important that access hatches and panels are available to gain access to the equipment being commissioned.

Throughout the commissioning phase all measurements and details should be recorded as the commissioning proceeds and submitted as commissioning records.

The commissioning of build services on major projects is often underestimated and the standard of building completion required to carry out this work has to be as near the finished project as possible.

Commissioning throughout should be carried out in accordance with the Chartered Institute of Building Services Commissioning Codes.

Electrical Installations

Following the satisfactory conclusion of inspection and tests on each completed section of the work commissioning can commence.

However prior to beginning the commissioning process a commissioning schedule should be produced. It should detail:

- ▲ plant to be commissioned
- ▲ operations to be carried out

- ▲ time scale
- ▲ exact dates for specific operations
- ▲ details of the requirements for water and power.

Commissioning should include:

- ▲ energising of electrical distribution circuits and equipment
- ▲ setting of electrical protective devices and systems
- ▲ starting up of all electrically powered plant and equipment
- ▲ confirmation of the performance of all plant and equipment by the carrying out of further tests and the making of all necessary adjustments to obtain optimum performance.

Mechanical Installations

A commissioning schedule providing the details as outlined for the electrical installation should also be produced to cover the commissioning of the mechanical installation.

Commissioning and performance testing of major items of equipment including boilers, air compressors, air handling plant etc should involve the manufacturer's personnel in conjunction with the commissioning engineer.

Commissioning should include:

- ▲ all air systems should be balanced
- ▲ all hydraulic systems should be balanced
- ▲ all distribution systems should be balanced and optimum noise levels reached
- ▲ all dampers should be clearly marked when the system has been balanced
- ▲ all valves should have their final positions recorded and should be locked in their final positions.

Public Health

The public health system should be tested in accordance with the local authority's requirements with either an air or water test.

Controls

Many installations now include BMS or Building Management Systems to control and monitor the entire operation of the services installation within the building. These systems are very specialised and require careful testing and commissioning.

Generally the BMS System cannot be set into operation and commissioned until the other commissioning operations have been completed.

It is vital that the decision to utilise a BMS is taken at the design stage. Attention should be paid to all aspects of the BMS to ensure its total integration into the construction process. The release of information must be as early as possible and should be issued to the mechanical and electrical contractors for detailed planning.

Lifts

As with all other service installations, the lift installation should be checked as the works proceed and commissioned in accordance with the requirements of the relevant British Standard.

Test results should be tabulated and submitted as records, and additional tests may be required for insurance certification.

FINE TUNING

Fine tuning of the entire building services installation cannot take place until the building is operational and time must be allowed for this process to be completed. It is basically the final adjustments to the system to compensate for items such as heat gain from equipment etc.

CONCLUSION

If commissioning is to be successful it is vitally important that clear records are kept throughout and the sequence and methods agreed at contract commencement are maintained.

Record drawings of the 'as built' installation should always be produced identifying the various components of the installations and providing maintenance details. Consideration should be given to the involvement during the commissioning period of the building owners' maintenance department in order that they can become familiarised with the services installation.

It should also be noted that 'Practical Completion' of a project often cannot be achieved without the satisfactory commissioning of life protecting systems such as fire alarms, safety shutters, gas and other extinguishing systems. The importance of commissioning building services is therefore at the forefront of the successful completion of major construction projects.

MEMBERSHIP OF THE MANAGEMENT OF MAJOR PROJECTS EDITORIAL GROUP

Chairman

Mr Christopher Tayler
Engineering Manager
Shell (UK) Ltd

Members

Mr Alan Hadden
Executive Councilman
GMB

Mr John Ling
Ling Management Consultants

Mr Ian McAlpine
Director
Sir Robert McAlpine

Mr Brian Meekey
Director of Operations
Fluor Daniel Ltd

Mr Robert Williams
Powergen

Mr Ivor Williams
Industrial Adviser
National Economic Development Office

MEMBERSHIP OF THE NEDC CONSTRUCTION INDUSTRY SECTOR GROUP

Chairman

Sir Christopher Foster
Director
Coopers & Lybrand Deloitte

Members

Mr Christian Adams
Head, Projects and Export Policy Division
Department of Trade and Industry

Mr Noel Bailey
Director
N G Bailey Organisation Ltd

Mr Tony Budge
Chairman
A F Budge (Contractors) Ltd

Mr David Cawthra
Chief Executive
Balfour Beatty Ltd

Mr David Compston
Chairman
Allott & Lomax Consulting Engineers

Mike Cottell
County Surveyor
Kent County Council
Highways and Transportation

Mr Barry Crowder
Director
Turbine Generator Services
NEI Parsons Ltd

Mr David Davies
Chairman
D Y Davies Associates Ltd

Sir Andrew Derbyshire
President
RMJM Ltd

Mr Roger Dobson
Director General/Secretary
Institution of Civil Engineers

Mr Alistair Fleming
Managing Director Construction
Eurotunnel Project Implementation Division

Mr Paul Gallagher
President
Electrical, Electronic, Telecommunications & Plumbing Union

Mr Christopher Groome
Head of Electronics, Engineering and Construction Section
National Economic Development Office

Mr George Henderson
National Secretary
Building & Construction Group
Transport & General Workers Union

Sir Brian Hill
Executive Chairman
Higgs & Hill plc

Mr David Holmes
Deputy Secretary of Highway Safety & Traffic
Department of Transport

Ian Howatt
Chairman of the Management Board
Franklin & Andrews

Mr Martin Laing
Chairman
John Laing plc

Mr Tom MacLean
General Secretary
Construction Section
Amalgamated Engineering Union

Mr Brian Moss
Chairman
NuAire Ltd

Mr Harry Nobbs
Chairman
Boon Edam Ltd

Mr Norman Nolan
Chief General Manager Europe
Pioneer International Ltd

Mr Patrick de Pelet
Director of Projects
Kleinwort Benson Ltd

Mr Tom Paul
Director and General Manager
Ward Building Systems Ltd

Mr David Plant
National Officer
GMB

Mr David Revolta
Head of IAE2
HM Treasury

Mr Peter Rogers
Director
Stanhope Properties plc

Mr Richard Rooley
Partner
Donald Smith, Seymour & Rooley

Mr Sandy Scott
National Industrial Officer
GMB

Mr Keith Sneddon
National Officer
Manufacturing Science & Finance Union

Mr Christopher Spackman
Chairman & Managing Director
Bovis Construction Ltd

Mr Christopher Tayler
Engineering Manager
Shell (UK) Ltd

Mr John Thane
Head of Research
National & Local Government Officers' Association

Mr Albert Williams
General Secretary
Union of Construction, Allied Trades & Technicians

Secretary

Mrs Beryl Garcka
Industrial Adviser
National Economic Development Office

Printed in the United Kingdom for HMSO
Dd 291775 C20 6/91